Energy Efficiency

edited by
Jenny Palm

Energy Efficiency

Edited by Jenny Palm

This is a reprint of the book published in 2010 by Sciyo

Republished by InTech

Janeza Trdine 9, 51000 Rijeka, Croatia

Publishing Process Manager Ana Nikolic

Technical Editor Sonja Mujacic

Cover Designer Martina Sirotic

Image Copyright Alfgar, 2010. Used under license from Shutterstock.com

Additional hard copies can be obtained from orders@intechopen.com

Energy Efficiency, Edited by Jenny Palm
p. cm.
ISBN 978-953-307-137-4

For EU product safety concerns:
IN TECH d.o.o., Prolaz Marije Krucifikse Kozulić 3, 51000 Rijeka, Croatia,
info@intechopen.com or visit our website at intechopen.com.

Contents

Preface

Global warming resulting from the use of fossil fuels is threatening the environment and energy efficiency is one of the most important ways to reduce this threat. Industry, transport and buildings are all high energy-using sectors in the world and even in the most technologically optimistic perspectives energy use is projected to increase in the next 50 years. How and when energy is used determines society's ability to create long-term sustainable energy systems. This is why this book, focusing on energy efficiency in these sectors and from different perspectives, is sharp and also important for keeping a well-founded discussion on the subject.

Transforming energy systems toward greater sustainability requires technological shifts as well as transformations in behaviour, values, and routines to conserve energy. This transformation can be facilitated by policy means and government initiatives as well as technological improvements and innovations. This book combines engineering and social science approaches to enhance our understanding of energy efficiency and broaden our perspective on policy making regarding energy efficiency. The book will be an essential read for anyone interested in how to contribute to the development of sustainable energy policies and achieve improved energy efficiency in industry, transport and the built environment.

The book is organised as follows. In the first chapter Morvaj and Bukarica discuss how to design, implement and evaluate energy efficient policy. This is followed by chapter 2 where Basosi and Ruzzenenti highlight the rebound effect and problematise why the world sees a growth in energy consumption despite the trend of higher efficiency.

The following three chapters focus on industrial energy efficiency. Thollander, Palm and Rohdin discuss earlier studies on industrial barriers and how STS-perspective can contribute to the barrier literature. Martinez compares factors that influence energy efficiency in German and Colombian manufacturing. Such comparison is important to improve our understanding of which factors are globally valid and which factors are more locally anchored. In chapter 5 von Schéele shows how specific technologies become important for achieving increased energy efficiency in industrial processes.

Chapters 6 and 7 in different ways relate to development in the building sector. In chapter 6 Rohdin, Glad and Palm have done a literature review on methods and main results in scientific publications on low-energy buildings and low-energy architecture. In chapter 7 Desha and Hargroves discuss education of built professionals, such as architects, planners and engineers, and the challenge and opportunities that exist for future professionals with extensive knowledge about energy efficiency in buildings.

The last two chapters both concern how different technologies can contribute to achieve ambitious policy goals on energy efficiency. In chapter 8 Jensen, Li and Bjerrum compare different hydrogen storage techniques in terms of energy efficiency and capacity available. In the last chapter Simeone, Salemme and Menna present a comprehensive analysis of energy efficiency of fuel processor.

Sustainable development demands new strategies, solutions, and policy-making approaches. This book discusses a wide spectrum of research on how to achieve ambitious policy goals on energy efficiency ranging from how energy efficient policy can be improved to how different technologies can contribute to a more energy efficient future.

Editor

Jenny Palm
Tema T, Linköping University,
Sweden

1

Energy efficiency policy

Zoran Morvaj[1] and Vesna Bukarica[2]
[1]*United Nations Development Programme (UNDP)*
[2]*University of Zagreb Faculty of Electrical Engineering and Computing*
Croatia

1. Introduction

Access to all forms of energy at affordable prices is an impetus for economic and social development of the society. At the same time, energy sector is responsible for approximately 75 percent of total greenhouse gases emissions, which makes it the main provocative of climate change. The convergence of international concerns about climate change and energy security in the past decade has led to the increased awareness of policy-makers and general public about energy issues and creation of new energy paradigm, the focus of which is **energy efficiency**. Energy not used is arguably the best, the cheapest and the least environmentally damaging source of energy supply and nowadays the concept of "negawatts" in energy strategies worldwide is being introduced. However, energy efficiency being typically demand side option is hard to implement due to the variety of stakeholders, i.e. players in the energy efficiency market that need to be stimulated to adopt energy efficiency as a way of doing business and ultimately a way of living - the change of mindset is needed. As higher efficiency of energy use is indisputably a public interest, especially in the light of the climate change combat, policy interventions are necessary to remove existing market barriers hindering the fulfilment of potentials for cost-effective efficiency improvements. Policy instruments to enhance energy efficiency improvements must stimulate the transformation of the market towards higher efficiency, with the final aim of achieving cleaner environment, better standard of living, more competitive industry and improved security of energy supply. Moreover, they have to be designed according to the real needs of the market (*tailor-made*), and have to have the flexibility and ability to respond (*adapt*) to the changing market requirements in order to achieve goals in the optimal manner.

Although there are excellent policies in place worldwide, with the European Union (EU) being the indisputable energy efficiency and climate change combat leader, the results in terms of reduced energy consumption are missing in the desired extent. Therefore, energy efficiency policy making needs new, innovative approaches the main feature of which is dynamics. Dynamic policy making means that it has to be learning, continuous, closed-loop process which involves and balances policy design, implementation and evaluation. The aim of this chapter is to explain these three main pillars of effective energy efficiency policy making, focusing especially on implementation issues, which are usually highly neglected in policy making process but are crucial for policy success.

2. Understanding energy efficiency policy making

2.1. Energy efficiency concept: avoid, reduce, monitor and manage

The basis for understanding the concept of energy efficiency is energy flow, from primary energy contained in energy carriers to the useful energy consumed through various activities of the society (Fig. 1).

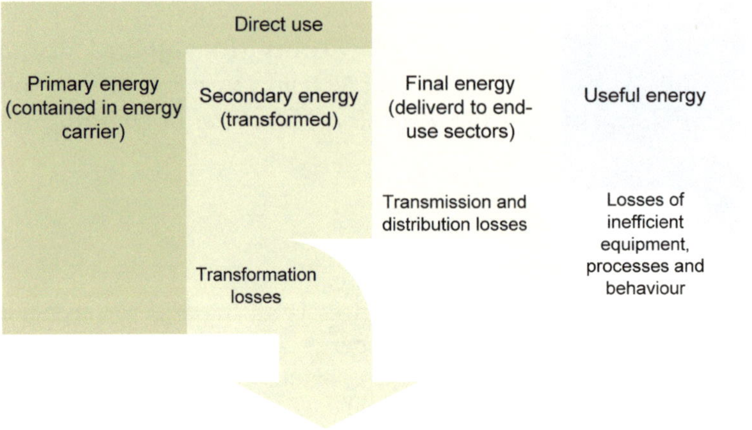

Fig. 1. Energy flow - basis for understanding energy efficiency

Energy efficiency is all about tackling energy losses. As shown in Fig. 1, it boils down to the very simple and understandable equation:

$$E_{useful} = E_{primary} - E_{losses} \tag{1}$$

Losses occur in processes of energy transformation, transmission, and distribution as well as in the final uses of energy. While reducing losses in the first three activities is mainly a matter of technology, the latest should be tackled by both technical and non-technical measures. Often unnecessary uses of energy could be avoided by better organisation, better energy management and changes in consumers' behaviour and increasingly so by changing lifestyle, which is the most difficult part. Energy efficiency has to be considered as a continuous process that does not include only one-time actions to avoid excessive use of energy and to minimise energy losses, but also includes monitoring and controlling energy consumption with the aim of achieving continuous minimal energy consumption level. Therefore, energy efficiency improvements rest on the following pillars (Morvaj & Bukarica, 2010):

- **Avoiding** excessive and unnecessary use of energy through regulation (e.g. building codes and minimal standards) and policies that stimulate behavioural changes;
- **Reducing** energy losses by implementing energy efficiency improvement measures and new technologies (e.g. waste heat recovery or use of LED lighting);
- **Monitoring** energy consumption in order to improve knowledge on energy consumption patterns and their consequences (e.g. smart metering and real-time pricing).
- **Managing** energy consumption by improving operational and maintenance practices.

To ensure continuity of energy efficiency improvements, energy consumption has to be managed as any other activity. Actually, energy management can be denoted as a framework for ensuring continuous avoidance of excessive energy use and reduction of energy losses supported by a body of knowledge and adequate measuring and ICT technology (Morvaj & Gvozdenac, 2008). It should not only consider techno-economic features of energy consumption but should make energy efficiency an ongoing social process. It also rests on the fact that energy has to be priced in a manner that more accurately reflects its actual costs, which include, inter alia impacts on the environment, health and geopolitics, and that consumers have to be made aware of these consequences of energy use. These main pillars for achieving energy efficiency improvements have to be taken into account in the policy making process - "avoiding" and stimulation of "reducing" shall be a main driver in design of policy instruments, while for "monitoring" and "managing" implementing capacities with appropriate capabilities and supporting infrastructure shall be ensured.

2.2. Rationale behind energy efficiency: means not an end

Energy efficiency shall be regarded as a mean to achieve overall efficient resource allocation (Dennis, 2006), rather then the goal in it self. As a consequence of improved energy efficiency, other public policy goals will be achieved as well, the most important of which are the goals of economic development and climate change mitigation.

In economic terms, and taking into account the fact that energy costs typically account to 15 to 20 percent of national gross domestic product, the significance of energy efficiency is evident - reduced energy consumption lowers the costs for energy. For example, it is estimated that the EU, although the world's most energy efficient region, still uses 20 percent more energy than it would be economically justified, which is the equivalent to some of 390 Mtoe (European Commission, 2006) or the gross inland consumption of Germany and Sweden together (Eurostat, 2009).

Furthermore, global consensus is emerging about consequences of inaction for mitigation of an adaptation to climate change, and clear quantifiable targets (limiting CO_2 concentration and temperature increase) within the given time frame (until 2012, than 2020 and finally 2050) need to be achieved if wish to avert a major disasters in the foreseeable future. For the first time energy policy making is faced with such strict constraints, which require a radically different approach in the whole cycle of policy making with special emphasis on policy implementation. Energy efficiency is globally considered to be the most readily available and rapid way to achieve desired greenhouse gases reductions in the short to medium term. And taking into account the possible grave threats of climate change, the time scale in energy policy has never been more important.

Let us briefly look at the evolution of energy policy making and the role of energy efficiency (Fig. 2.). The standard energy policy making approach implied balancing of energy demand and supply and slow evolution of policy goals, mixes and objectives as a response to various external changes and drivers. The standard energy policy making was not faced with serious constrains and specifically not time constraints for achieving certain results and objectives. The time scales of energy policies were rather long, actions were gradually undertaken (leading often to under investing in energy sector) and mainly left to the decisions of energy companies, which led to the critical neglect of energy policy implementing capacities at various levels of jurisdiction and in the society in general.

Nowadays, energy policy is entering a new constrained phase, with time as the main constrain being imposed by the desire to combat climate change.

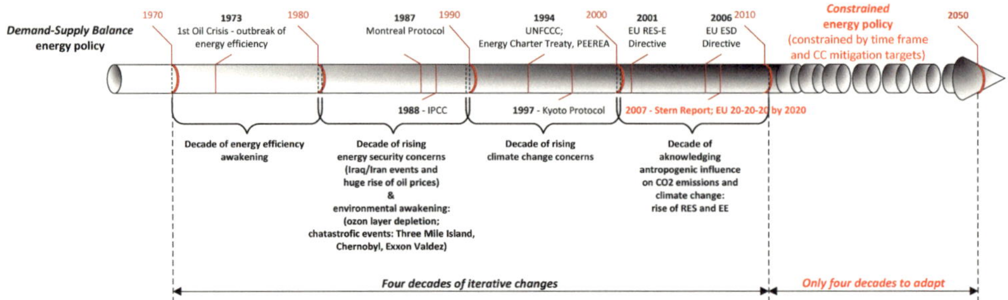

Fig. 2. Gradual changes of energy policy accents due to various drivers (Morvaj & Bukarica, 2010)

Energy efficiency solely can deliver the desired greenhouse gases reduction targets to the large extent. To confirm the statement, the EU has been taken as an example. It is estimated that fulfilling 20 percent target for energy efficiency improvements by 2020 would mean reducing greenhouse gases emissions by 780 million tonnes, more than twice the EU reductions needed under the Kyoto Protocol by 2012 (European Commission, 2006). Since the EU has committed to reduce its greenhouse gases emissions by 20 percent compared to 1990 by 2020 and since the EU's greenhouse gases emissions in 1990 amounted 5,564 million tonnes (European Environment Agency, 2009), it is evident that 20 percent of energy efficiency improvement can deliver almost three fourths of desired greenhouse gases reduction target. The power of energy efficiency as a tool for climate change combat is therefore obvious.

2.3. Levels of energy efficiency policy: from enabling to implementing

Taking into account the role energy efficiency plays in reaching global goals of climate change combat, it is understandable that there is a need for coordinated actions at all levels - international, regional (e.g. European Union) and national to ensure enabling environment for energy efficiency improvements by formulating appropriate policy instruments. However, the real power to change is local. Policies have to be designed in a way that enables local implementation in homes, public services and businesses. The interconnection between levels of energy efficiency policy is illustrated in Fig. 3.

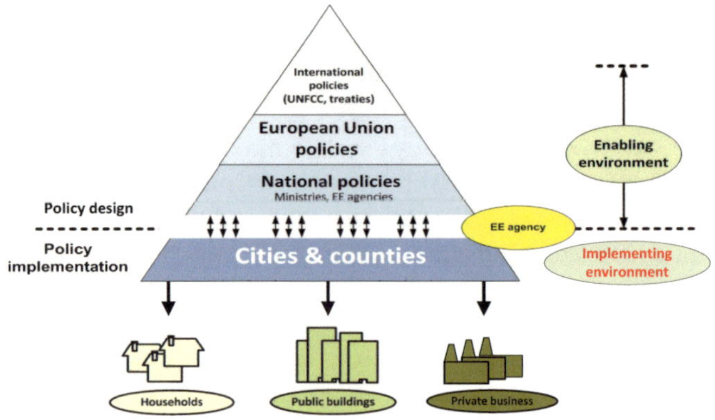

Fig. 3. Levels of energy efficiency policy

2.3.1. International aspect of energy efficiency policy

Due to its significance, energy efficiency is the topic of international agreements related to climate change combat, environmental protection and security of energy supply. Money and effort are put into promotion of energy efficiency by numerous international institutions, as briefly demonstrated in Table 1.

International treaties and agreements on Climate Change and EE		
Name of the document	***Year***	***Main features***
Energy Charter Treaty	1994	Legally-biding multilateral instrument, obliging parties, *inter alia*, to reducing negative environmental impact of energy cycle through improving energy efficiency
Energy Charter Protocol on EE and Related Environmental Aspects (PEEREA)	1994	Recognises EE as considerable source of energy and obliges parties to promote EE and to create framework which will induce both producers and consumers to use energy in the most efficient and environment friendly way as possible
Kyoto Protocol to United Nations Framework Convention on Climate Change (UNFCCC)	1997	Obliges parties to reduce GHG in time period 2008-2012. Defines flexible mechanisms that will ease the achievement of targets at the least cost
International institutions/programmes for energy efficiency		
Institution/Programme	***Year***	***Main features***
Global Environment Facility	1991 - 2009	GEF is main financial mechanism of UNFCCC; GEF has supported 131 EE projects with portfolio of approximately 850 million USD
World Bank Group	2005-2009	Renewable energy and EE at the heart of WBG energy agenda; in period 2005-2009 over 4 billion USD given for EE projects world wide
United Nations Development Programme, United Nations Foundation	/	Energy as an important factor in reaching Millennium Development Goals and reducing Poverty; Calls for international "Efficiency First" agreement; Number of EE projects financed world wide
International Energy Agency	/	EE one of six broad focus areas of IEA's G8 Gleneagles Programme - IEA submitted 25 policy recommendations to the G8 for promoting EE that could reduce global CO2 emissions by 8.2 gigatonnes by 2030.

Table 1. International treaties and programmes for energy efficiency (Morvaj & Bukarica, 2010)

As seen from Table 1, international treaties and programmes are supported by various financing tools, bilateral and international donors, but there is very little focus on how to implement policy measures and instruments, hence the real results in terms of sustainable and verifiable energy efficiency improvements and greenhouse gases reductions are missing. It is absolutely crucial to shift the focus of international policies towards real-life application, respecting in this process different local circumstances.

Namely, the drivers for energy efficiency and implementing environments differ significantly on the global scene. Four "blocks" could be identified as shown in the Fig. 4. The EU, followed by some other OECD countries, is certainly a forerunner in combating climate change and in related energy efficiency activities. USA and BRIC countries are the most vocal in defending their national interests and resisting any firm commitments for CO_2 reduction. Developing countries collectively represent a significant block in terms of greenhouse gases emissions. Energy efficiency is for them a win-win approach for reducing the greenhouse gases emissions while also reducing costs of energy for their fragile economies. Therefore, energy efficiency in developing countries should be addressed immediately and incorporated in energy policies with strong supporting implementation mechanisms.

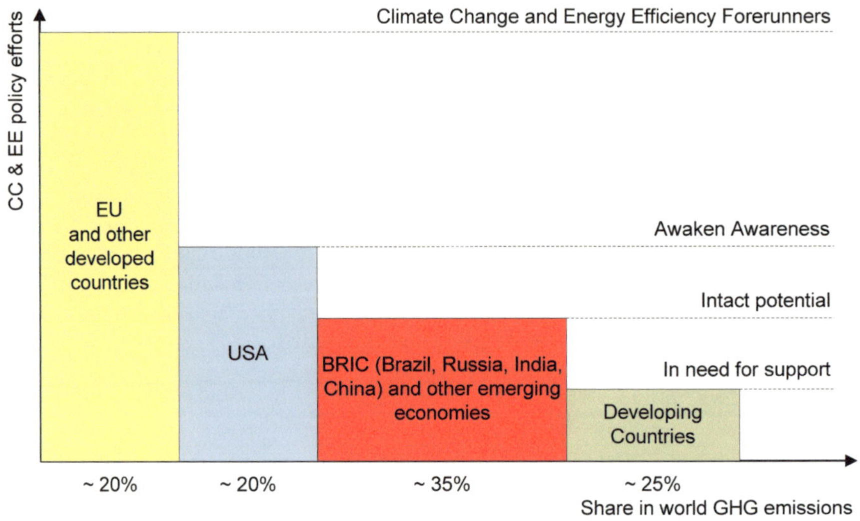

Fig. 4. World differences in climate change and energy efficiency policies adoption (Morvaj & Bukarica, 2010)

The efforts from the international level are extremely useful and necessary, but they are still not enough, i.e. they are generic in their nature, hence are not able to deliver real results. International policies, programmes and aids shall be brought down to the national and local level in every "block", where conditions for policy implementation are different, requiring thus tailor-made solutions in both policy instruments and implementing capacities.

2.3.2. Regional energy efficiency policy: case EU

The indisputable "energy efficiency forerunner" in the world is the European Union (EU). The EU has strongly stressed its aim to achieve the "20-20-20" targets by 2020: to reduce greenhouse gases emissions minimally 20 percent (with the intention to even achieve 30 percent greenhouse gases emission cut by 2030); to increase the proportion of renewable energies in the energy mix by 20 percent and to reduce primary energy consumption by 20 percent. In order to achieve the energy efficiency improvement goals, the EU has introduced a well thought of set of voluntary and some mandatory polices. The most important policy and legislative documents related to energy efficiency in the EU are summarised in the Table 2.

EU policy documents on EE		
Name of the document	***Year***	***Main features***
EE in European Community – Towards a Strategy for the	1998	Analyse available economical potential for improvements in energy efficiency, identifies barriers

Rational Use of Energy (COM (1998)) 246 final)		and gives proposals to remove those barriers. Estimates that saving of 18% of 1995 energy consumption can be achieved by 2010 (160 Mtoe).
Action Plan to Improve EE in the European Community (COM (2000) 247 final)	2000	Sets a target for energy intensity improvement by an additional 1% per year compared to a business as usual trend resulting in 100 Mtoe avoided energy consumption by 2010.
Green Paper on EE or Doing More with Less (COM (2005) 265 final)	2005	Expresses urging need to put energy saving policy higher on the EU agenda and estimates that EU is using 20% more energy then economically justifiable and if additional efforts are not made, this potential will not be fulfilled by current policies.
Action Plan for Energy Efficiency: Realising the Potential (COM(2006) 545)	2006	Sets energy saving target of 20 percent by 2020 (390 Mtoe) and defines 6 priority policy measures (energy performance standards; improving energy transformation; focusing on transport; providing financial incentives and ensuring correct energy pricing; changing energy behaviour; fostering international partnership).
Second Strategic Energy Review - An EU Energy Security and Solidarity Action Plan (COM/2008/0781)	2008	Reinforces EE efforts to achieve 20% target - calls for revision of directives on energy performance of buildings, appliance labelling and eco-design, strongly promotes Covenant of Mayors, use of cohesion policy and funds and tax system to boost energy efficiency.
EU EE legislation (directives)		
Directive 92/75/EEC on energy labelling of household appliances and implementing directives	1992	Prescribes obligatory EE labelling for 8 groups of household appliances.
Directive 2002/91/EC on the energy performance of buildings (Proposal for a Directive on the energy performance of buildings (recast) [COM(2008)780])	2002 (recast proposed in 2008)	Calls for minimum energy requirements for new and existing buildings, energy certification and regular inspection of boilers and air conditioning systems.
Directive 2004/8/EC on the promotion of cogeneration based on a useful heat demand in the internal energy market	2004	Facilitate the installation and operation of electrical cogeneration plants.
Directive 2005/32/EC establishing a framework for the setting of eco-design requirements for energy-using products and implementing directives	2005	Defines the principles, conditions and criteria for setting environmental requirements for energy-using appliances.
Directive 2006/32/EC on Energy end-use Efficiency and Energy Services	2006	Calls for establishment of indicative energy savings target for the Member States, obligations on national public authorities as regards energy savings and energy efficient procurement, and measures to promote EE and energy services.

Table 2. EU policy documents for energy efficiency (Morvaj & Bukarica, 2010)

The analysis of these documents clearly shows the commitment and huge policy efforts to boost energy efficiency improvements. Despite that, the EU is far from reaching its 20 percent energy efficiency improvement target by 2020. The results of the policy implementation are missing in the desired extent, leaving the huge potential of "negawatts" idle. With the current legislation and policy instruments in place, a reduction of only 8.5 percent will be achieved. Even taking into account additional measures in the pipeline, at the best only 11 percent reductions will be achieved, as shown in the Fig. 5 (European Commission, 2009). However, the EU policy only provides the framework national policies have to cope with. It is, to the largest extent, the task of national policies to deliver actual energy efficiency improvements. Obviously, they are failing to do so.

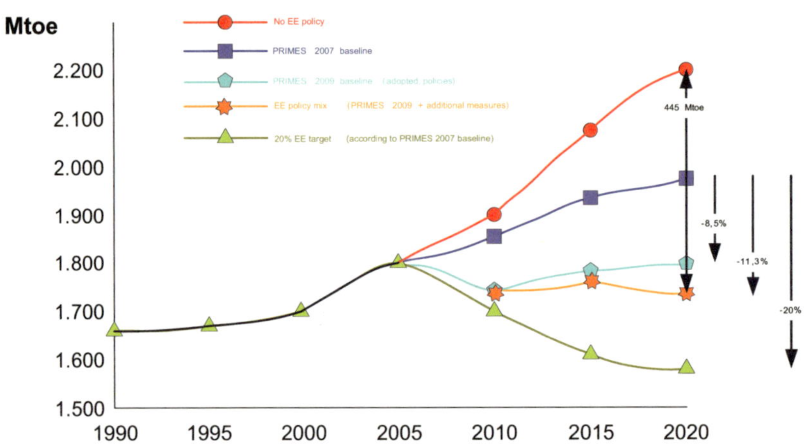

Fig. 5. Development and projection of Gross Inland Energy Consumption for EU by 2020 (European Commission, 2009)

2.3.2. National energy efficiency policy: (not) delivering targets

In national energy efficiency policy there is a symptomatic unbalance between efforts for preparing polices, and preparations for policy implementation. The vast majority of policy makers are focused on incorporating requirements of international policies and requirements into national strategic and legislative frameworks, without thorough consideration of national circumstances, i.e. without taking into account the level of energy efficiency market maturity in a country. Moreover, there is a general lack of focus on policy implementation and a sort of general expectation that implementation is straightforward, will hopefully happened by itself, hence there is no need to put too much efforts into that. Current national energy efficiency policies are persistently missing or underachieving the desired results. There are number of reasons behind this policy failure, but the problem is essentially threefold:

1. **Policy makers do not fully tackle all stakeholders relevant for energy efficiency,** i.e. not all market players are tackled with appropriate policy instruments that would remove market imperfections and enable sustainability. There is a need for all-a-compassing, tailor-made policies, adaptive to specific changing market conditions.

2. **Policy making needs to appreciate specific implementing environment conditions and time constraints for implementation**, thus focusing on creating sufficient and appropriate implementing capacities that are adequate for achieving the targets. A model for developing implementing capacities shall be established.
3. **Policies are not static**, meaning that policy making is not on-time job. It requires well established procedures for policy monitoring and evaluation that will reveal what works and what does not work in the practice and provide inputs for policy improved redesign.

Obviously, new approach in overall energy efficiency policy making is needed, the main feature of which is dynamics.

2.4. Policy dynamics: key to effective energy efficiency policy making

For energy efficiency policy to be successful its creation has to be a learning process based on both theoretical knowledge and empirical data. This learning process can be the most appropriately described by the closed-loop process (Fig. 6) consisting of the following stages:

- **Policy design**:
  - **Policy definition**: objectives, targets, approaches for different target groups, legal and regulatory frameworks;
  - **Policy instruments development**: incentives, penalties, standards, technical assistance, financing support;
- **Policy implementation**: institutional framework, stakeholders, human resources, capacity and capability development, supporting infrastructure (ICT);
- **Policy evaluation**: monitoring of achieved results through energy statistics and energy efficiency indicators, qualitative and quantitative evaluation of policy instruments' impacts.

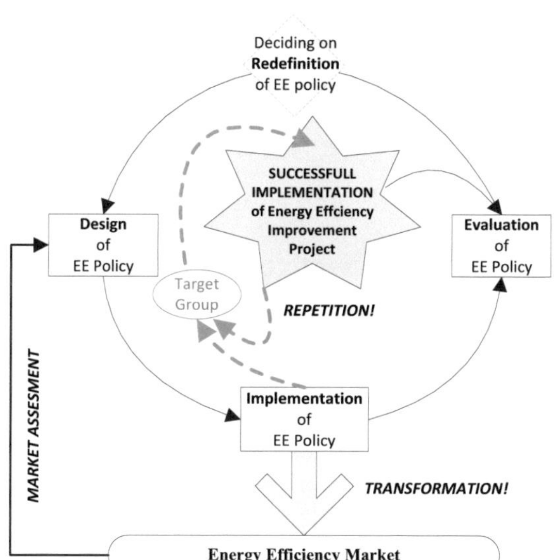

Fig. 6. Dynamics of energy efficiency policy (Bukarica et al., 2007)

Energy efficiency policy in its essence shall be a market transformation programme. Market transformation programmes are strategic interventions that cause lasting changes in the structure or function of markets for all energy-efficient products/services/practices (Brinner & Martinot, 2005). The effective market transformation programme rests on the following key pillars:

- mix of policy instruments created to **remove market barriers** identified throughout all stages of the individual energy efficiency project development;
- policy interventions **adaptive to market conditions** ensuring sustainability of energy efficiency improvements through replications of successfully implemented energy efficiency projects;
- **policy instruments tailored** to enable all market players (government, private sector, consumers, equipment producers, service providers, financing institutions, etc.) to find their interest in improved energy efficiency;
- energy efficiency improvements achieved as the result of **supply-demand interactions based on competitive market forces**.

Therefore, prior to the start of energy efficiency policy design the market assessment shall be preformed. It shall reveal the maturity of the market. This is extremely important, as different instruments have different effects and are therefore appropriate at different market maturity levels, i.e. some measures could stimulate market introduction, whereas other measures could accelerate commercialisation, or increase the overall penetration of energy-efficient products and services (Brinner & Martinot, 2005). Market analysis is required to identify market forces that have to be strengthened by incentives or diminished by penalties. The policy instruments should be carefully designed and mixed in order to tackle identified market barriers.

Conceptually, the typical energy efficiency policy cycle starts with strategic planning and determination of targets leading to the design of specific instruments to tackle different target groups, i.e. market players. The implementation of policy instruments follows and one cycle is concluded with the evaluation of policy impacts. The results of the policy evaluation process are then fed into the planning, design, and implementation processes, and the cycle repeats itself (Vine, 2008). Every stage in this dynamic loop requires methodical and systematic approach and will be given all due attention in the subsequent sections.

3. Main postulates for defining effective energy efficiency policy

3.1. Understanding energy efficiency markets

The starting point in creation of any policy is to understand how market operates and how well developed it really is. Unlike the economic theory that assumes perfect competition, the real markets are imperfect due to various barriers preventing market forces to deliver desired results. The task of any policy is to identify these barriers and to develop market-based incentives and well-designed, forward-looking instruments for their removal (Dennis, 2006).

Policies usually define various instruments to support implementation of energy efficiency measures in energy end-use sectors (households, services, industry, transport). Very often, the proposed instruments are generic and designed without a proper appreciation of the situation on the ground – an energy efficiency market place where energy efficiency measures need to be adopted by consumers, supported by energy service providing companies. Addressing end-users solely is not nearly sufficient to ensure self-sustainable energy efficiency improvements. The concept of energy efficiency market shall be introduced and understood for creating and implementing energy efficiency policy.

Energy efficiency market is not exactly one market but a conglomeration of various and very diverse businesses acting in the field and having different interests in energy efficiency realm. Energy efficiency market's supply side includes providers of energy efficient equipment and services as well as institutions involved in financing and implementation of energy efficiency projects (banks, investment funds, design engineers, constructors, etc.). The demand side of energy efficiency market includes project sponsors with ideas for energy efficiency improvements (end-users, i.e. building owners and renters, building managers, public sector institutions and local authorities, industries).

The performance of energy efficiency market is evaluated according to the actual energy efficiency improvements delivered, i.e. according to number of successfully implemented energy efficiency projects. Basically, the energy efficiency market transformation depends on the success of the project development process. Development of an energy efficiency project goes through various stages, from the very initial idea, until the final and actual implementation of the project that operates and yields results in terms of reduced energy consumption and emissions (Fig. 7). Due to various market barriers, only few of a variety of identified opportunities for energy efficiency improvements reach the stage of a bankable project, becoming actually implemented; hence the narrowed pipeline presentation is chosen.

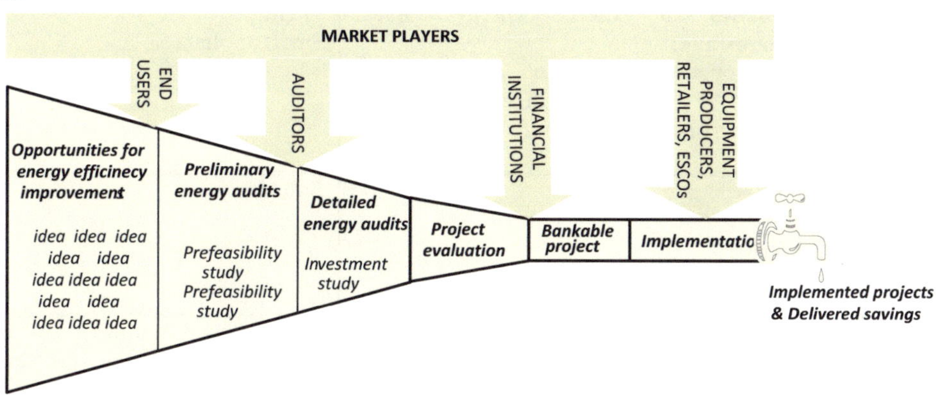

Fig. 7. Understanding energy efficiency projects' development cycle and energy efficiency markets (Bukarica et al., 2007)

3.2. Definition of policy instruments for market transformation

One of the main reasons for energy efficiency policy failure lies in the preference of policy makers to use universal solutions in definition of energy efficiency policy and basically to copy-paste policy instruments from others without considering the specificities of own country's energy efficiency market. There are, of course, some general market barriers for energy efficiency which require such universal solutions (Table 3), but they are not nearly sufficient to provoke market transformation and to fulfil the final goal - creation of self-sustainable energy efficiency market.

Primary Barriers	Effects	Solutions
Incomplete (imperfect) information	Affects both demand and supply side of EE market leaving the demand underdeveloped and supply side disinterested	Dedicated promotional and informational campaigns; Energy labelling of appliance, equipment, buildings and cars
EE as public goods	Markets tend to undersupply public goods	Stimulating Research and Development of energy efficient technologies; Voluntary agreements with manufacturing industries
Externalities	Energy price does not reflect the adverse environmental and human health effects of energy consumption nor impacts of political instabilities related to energy supply; Positive externalities of improved EE should also be taken into account.	Correct energy pricing and energy taxation; Environmental fees (but usually imposed to large consumers only); Tax credits for EE investments ; Minimal efficiency standards; Utilising purchasing power (green public procurement and consumers' awareness)
Market power (imperfect market structures)	Remains of monopoly in energy sectors prevent development of truly competitive energy markets and restructuring of utilities to become energy service companies; Improper structures of energy prices based on historical average costs and not on short-run marginal costs	Transforming utilities to become energy service companies; Smart metering and real-time pricing; Smart appliances
Secondary Barriers - consequences of primary barriers	**Effects**	**Solutions**
Lack of access to capital	Makes it difficult or impossible to invest in energy efficiency	EE (revolving) funds (as initial driver of demand for energy efficient solutions); Transforming utilities to become energy service companies
Mindset (rather then market) barrier	**Effects**	**Solutions**
Consumers' behaviour	Optimal decisions will not be made regardless sufficient information provided due to bounded rationality	Energy and climate literacy (a top educational priority in schools and in the public discourse)

Table 3. General market barriers to energy efficiency and universal solutions (Morvaj & Bukarica, 2010)

Instead of routine proposals of generic policy instruments, specific status of energy efficiency market in a given jurisdiction has to be understood, and for every stage in the energy efficiency project development process specific barriers must be identified and support policy instruments designed to ensure project pipeline throughput (Bukarica et al. 2007). In other words, policy instruments have to be **tailor-made** for specific market circumstances.

Energy efficiency market has a variety of players with different backgrounds and as such is highly influenced by behavioural, socio-economic and psychological factors that govern market players' decisions. All these influences have to be taken into account when defining policy instruments for energy efficiency improvement. As indicated in the Fig. 8, combination of policy instruments has to be used to remove both supply and demand side barriers, i.e. both supply and demand side have to be addressed simultaneously when markets are "stuck". In other words, producers/service providers have to be stimulated to produce/offer more efficient products/services, while consumers have to be stimulated to by such products/services. What this means is that if there is no demand for energy efficient products/services suppliers are not interested in improving their performance by themselves and vice verso, if there is no efficient products/services offered in the market, there is no demand for them either. Policy instruments have to be designed to move this situation from the deadlock and to fulfil the ultimate goal of market transformation - to achieve public benefits from increased energy efficiency as accepted mode of behaviour (Bukarica et al., 2007).

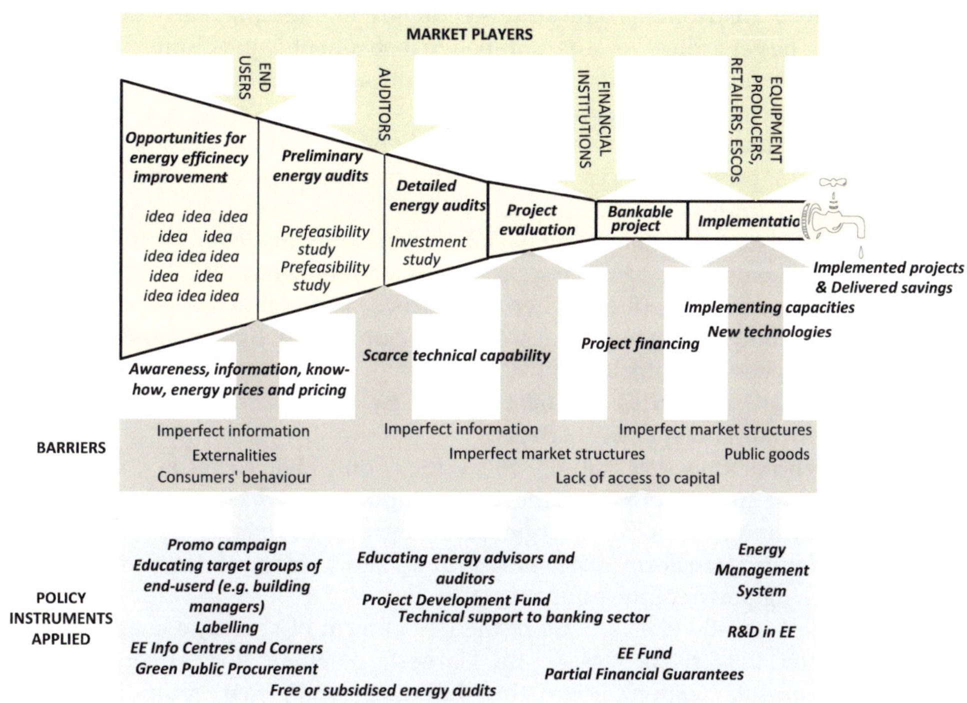

Fig. 8. Defining energy efficiency policy instruments based on actual status of a specific energy efficiency market (Morvaj & Bukarica, 2010) (*Note: the scheme was developed during market assessment and creation of energy efficiency policy in the Republic of Croatia*)

Policy-makers have to understand that policy instruments are not equally relevant at all points in time - the requirement for different instruments vary with maturity of the market and timing of utilisation. Therefore, policies have to be **adaptive** to changing market conditions.

Adaptive policy response means that utilisation of instruments and funding designated for their implementation must correspond to the market demands. E.g. offering partial financial guarantees to the banks will have very modest impact in markets where there is no demand for energy efficiency projects and banks do not find the interest to offer specialised financial products for the. As a general guideline, instruments for awareness raising and technical assistance are more important in developing energy efficiency markets, while with its maturity financial incentives become increasingly desired.

- **Not all policy instruments are suitable for all markets**:
  - Understand the maturity level of country's energy efficiency market and tailor policy instruments to overcome identified barriers;
  - Use experiences of others, but do not copy-paste without taking into account real market situation - what works in one country, does not have to work in other;
  - Every policy instrument has its right timing for implementation - take one step at time to ensure smooth transformation of the market i.e. smooth transition from one phase to another as shown in Fig. 7;
- **Not all policy instruments are suitable for all market players** - be specific in determining target groups for a certain policy instrument (e.g. voluntary agreements are not suitable for households consumers, while appliance labelling will have little to do with large industry consumers);
- **Not all policy instruments are suitable for all energy end-use sectors** (households, public services, private services, industry, and transport) - sectors' specificities shall be taken into account;
- Sometimes it is useful to determine **package of instruments** (combinations of two or more instruments, e.g. building code in combination with subsidies for demonstrating achievement of higher standards or promotion campaign for cleaner transport in combination with subsidies for purchasing hybrid cars) to increase policy effectiveness and efficiency;
- **Identify sectors that can be the best tackled by policy** and that would have the largest immediate and spill-over effects:
  - Experience shows that putting policy focus on public sector is both easiest to implement and it provides the largest spill-over effect to other sectors by demonstrating effects of energy efficiency improvements, but it also has a potential to transform the market in a short span of time due to large purchasing power of the public sector;
  - Buildings usually consume more then 40 percent of country's energy demand, therefore this sector offers the largest potential for energy efficiency improvements (especially existing building stock) that could be achieved through advanced building codes and energy performance standards;
- **Look for local best practices and make them national** - often there are local initiatives in a country that have great results and capability for replication;
- **Be aware of your implementing capabilities** - available budget and, even more important, institutional capacities needed for implementation of policy instruments.

Box 1. "Quick-win" guidelines for designing successful energy efficiency policy instruments

4. Energy efficiency policy implementation

4.1. Understanding implementing environment

The immediate questions aimed at understanding the "implementing environment for energy efficiency policies" are:

- **Who has to do what?** In other words, what are the roles and responsibilities of different stakeholders.
- **Were the implementation has to happen?** The answer, although as simple as possible, is often overlooked - policy needs to be implemented where energy is used everyday – and this is at our places of work and at our homes.

It is very simple fact that all energy delivered is consumed directly by people or indirectly through different institutional and business forms created by people (Fig. 9), during the course of our professional and private life. Therefore, for implementation of energy efficiency measures and a full policy uptake, the mobilisation and cooperation of all stakeholders is needed. The international institutions and efforts form an umbrella of this implementing environment, dictating the framework for policy creation and implementation (as discussed in the section 2). At national level, four key groups of stakeholders, i.e. vertical social structures can be identified (Fig. 9), all of which have their specific roles in energy efficiency policy implementation and their activities (or lack thereof) influence the energy efficiency market.

The primary role of the **public sector institutions** is to ensure national policy implementation in all end-use sectors (households, services, industry and transport). However, at the same time the public sector, same as **businesses**, are the realms where policy is actually being implemented. **Civil society organisations** and **media**, on the other hand, play the key role in providing information and promoting energy efficiency on the wide scale, which will, in the long run, enable changing the consumers' mindset towards more energy efficient behaviour.

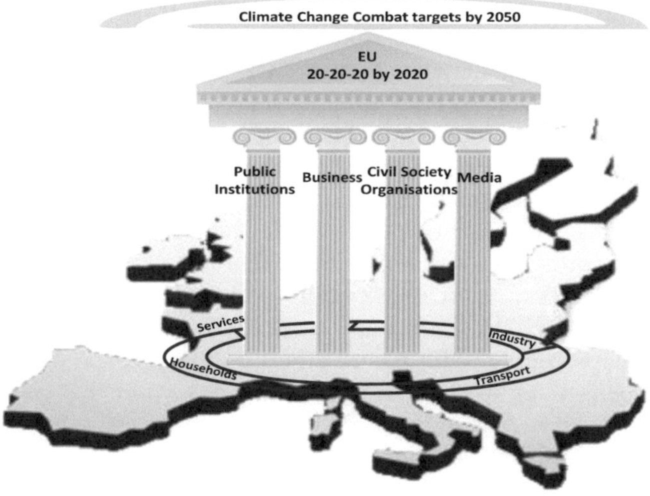

Fig. 9. Main pillars of implementing environment for energy efficiency policy

4.2. Roles and responsibilities of key stakeholders

Public institutions play, with no doubt, pivotal role in enabling and enhancing policy implementation. However, the governments, i.e. competent ministries themselves rarely have the capacities to deal with policy implementation issues. Therefore, in many countries specialised national energy efficiency agencies are established as governmental implementing bodies. They have a crucial role in initiating energy efficiency programmes, coordination of activities and especially in monitoring and evaluation of policy implementation.

To support this statement, a fact that nowadays more than 70 percent of European population lives in cities has to be emphasised. Even more so, in 2009 for the first time in history official statistics have reported that globally more than 50 percent of world population lives in cities. Hence cities are obvious places where vigorous, continuous and focused implementation of energy efficiency measures needs to be carried out by all key stakeholders (see Fig. 3).

Being closest to places where energy is consumed and still having executive powers, local authorities more than ever have a pivotal role to play at reducing energy consumption. Actions that local authorities (and public sector in general) should undertake are twofold:

- Firstly, energy consumption in facilities and services in their jurisdiction should be properly managed. This means that local authorities shall demonstrate their commitment by implementing energy efficiency improvement measures in all buildings in their jurisdiction (office buildings, schools and kindergartens, hospitals, etc.) as well as in public services they provide (public lighting, transport, energy and water supply).
- Secondly, information must be made publicly available and cooperation with civil society organisations, businesses and media has to be established to improve citizens' awareness and facilitate change of energy related behaviour and attitude.

Building local capacities to perform these activities is the most important precondition for successful policy implementation and delivering policy targets. Introduction of full-scale energy management is instrumental there, which could be a backbone for evolution of "smart cities" and sustainable urban development (Paskaleva, 2009).

In all business sectors, the climate change awareness and social responsibility are driving companies to demonstrate their "greenness". The new "green" revolution in the corporative world is led by the biggest - Google and Microsoft are going solar, Dell is committed to neutralising carbon impact of its operations, Wal-Mart aims at completely renewable energy supply, crating zero waste and selling products that sustain resources and the environment (Stanislaw, 2008). However, while corporations do have money and human capacities to turn their business towards more efficient and environmentally friendly solutions, small and medium enterprises (SMEs) need role-models and support to improve their energy efficiency, hence the overall business performance. The 2007 Observatory of EU SMEs indicates that only 29 percent of SMEs have instituted some measures for preserving energy and resources (46 percent in the case of large enterprises) and that only 4 percent of EU SMEs have a comprehensive system in place for energy efficiency, which is much lover then for large enterprises (19 percent) (European Commission, 2009). Again, energy management is the solution.

And finally, policy makers together with civil society organisations, businesses and media have to work together to ensure that energy and climate change literacy (Stanislaw, 2008) becomes a top educational priority in schools and in the public discourse. In this task, civil

society organisations and media have particularly important role, since they formulate the public opinion and are able to establish a new "green" ethic in rising generations.
Therefore, the solution for ensuring proper implementing environment for energy efficiency policies lies in bringing together and mobilizing for action all stakeholders so that every pillar of the society contributes fully according to their own means for achievement of energy efficiency policy targets. Strong links, as demonstrated in Fig.10, between each and every stakeholder shall be established, not only whilst implementing policy, but immediately during the process of energy efficiency policy design. Either link is equally important as the current practice has indicated that policy making lacking feedback from all stakeholders results in weak and slow implementation. The Fig. 10 aims to illustrate the need for stakeholders' interactions in various energy efficiency activities, and points that such coordinated and collaborative approach will influence citizens and eventually transform the market and society towards higher efficiency.

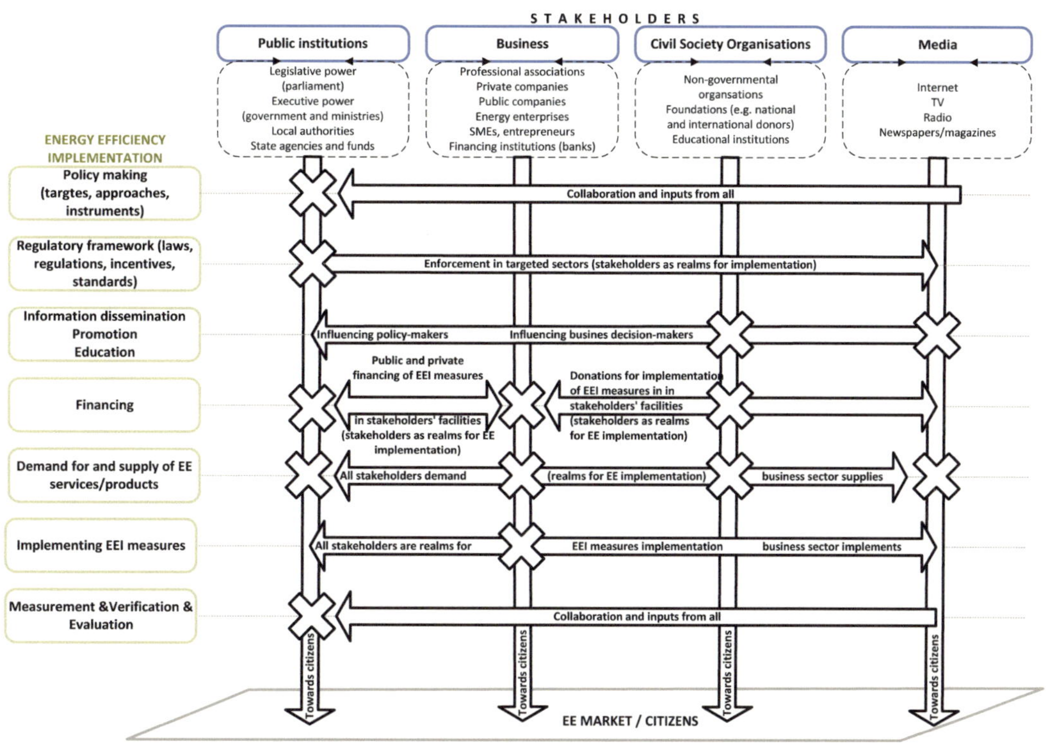

Fig. 10. Stakeholders' interactions in different energy efficiency activities

4.3 Building implementing capacities through Energy Management System

Implementing capacities can be successfully strengthen through the process known as Energy Management System (EMS). It comprises a specific set of knowledge and skills based on organizational structure incorporating the following elements:

- people with assigned responsibilities
- energy efficiency monitoring through calculation and analysis of:
    - energy consumption indicators
    - energy efficiency improvement targets

- continuous measuring and improvement of efficiency.

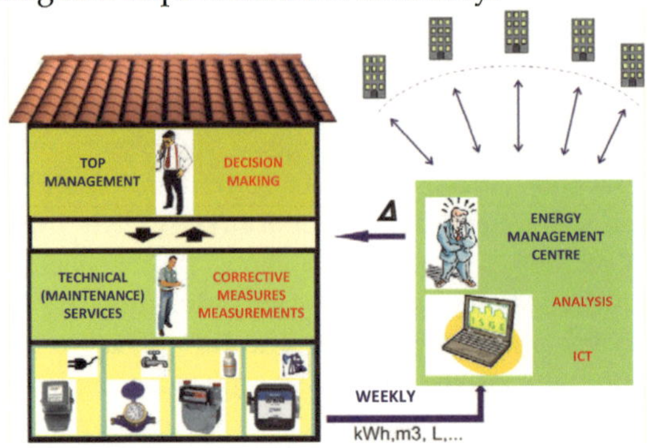

Fig. 11. Concept of energy management system (Note: EMS is equally applicable in public and business sector)

The process of introducing energy management starts from the decision of adopting an energy management policy statement. It then leads to an energy management action plan being adopted at the top management level. Measurable goals to be achieved are set within the plan. The plan with defined goals is made public. This act ensures a constant support of the top management and all employees to the implementation of energy management project. This is followed by introduction of organizational infrastructure to deliver the plan. A dedicated energy management team is appointed which assumes the obligation of overall energy management on the level of a city or a company. Furthermore, every facility in the structure of a company or in the ownership of a city has to have a person (usually technical or maintenance) appointed as the one responsible for the local energy management. And finally, all members of energy management team shall be adequately educated and trained to perform their tasks. This way capacities and capabilities for implementation of energy efficiency projects are ensured. Additionally, they need to be supported by appropriate ICT tool for continuous collection, storage, monitoring and analysis of data on energy consumption. Moreover, energy management team is also responsible for further educational and promotional activities to change employees' behaviour and attitudes towards energy consumption at the work place and for initiating green public procurement activities to stimulate market transformation by utilizing public sector's huge purchase power. And last, but not the least, energy management teams, especially those established within pubic sector (i.e. local authorities) are reaching out to the citizens by publicly announcing their activities and by providing advisory services. This comprehensive process of energy management system introduction is shown in the Fig. 12. Although it shows the process applied in the cities, it could be easily adjusted for business sector as well.

Once it is understood that policy implementation is happening locally, capacitating both public and commercial business market players for implementing energy efficiency policy through systematic introduction of energy management practices becomes the key to the policy success.

Another look at the Fig. 8 reminds us that implemented projects are only vehicle that deliver actual energy consumption reductions and they appear merely like a drops at the end of pipeline that involves huge number of actors, actions, barriers and instruments to overcome them. Without strong, focused, competent and effective capacities for implementing energy efficiency policies it is unlikely to expect that projects would flow from the pipeline and that the targets would be delivered.

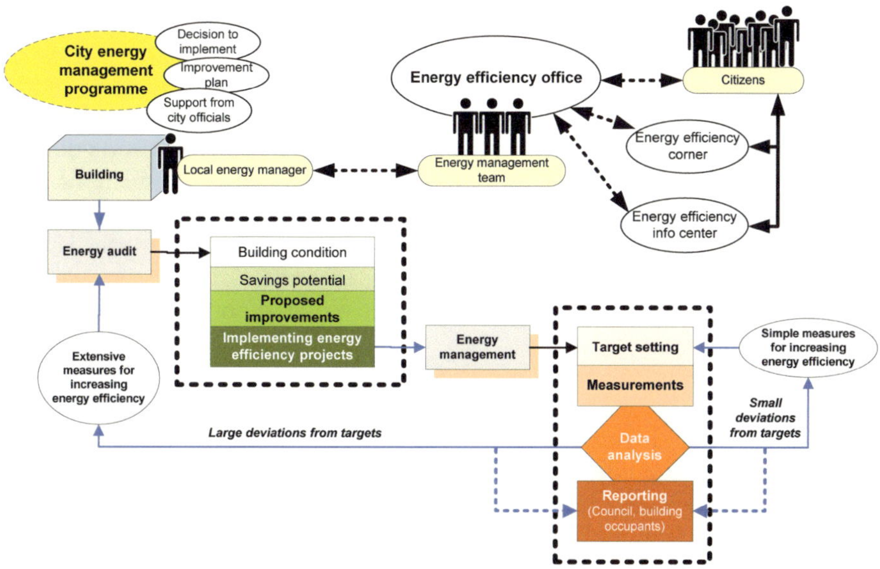

Fig. 12. Energy management process in a city (Note: The scheme is applied in the cities of the Republic of Croatia. The process is easily adjusted for business sector.) (Morvaj et al., 2008)

5. Evaluating energy efficiency policy: measurement and verification (M&V)

5.1. General issues on policy evaluation

In the energy efficiency policy cyclic loop policy evaluation has an essential position, although it might not appear so. Namely, evaluation procedures are at the same time an integral part of policy design phase as well as both parallel and consecutive activity to policy implementation.

The first step in policy design shall be establishment of a plausible theory on how a policy instrument (or a package of instruments) is expected to lead to energy efficiency improvements (Blumstein, 2000). Based on well-reasoned assumptions (theory) policy instruments mix shall be created. Well-reasoned means that strong believe exists that exactly this instrument will lead to cost-beneficial improvements in energy efficiency market performance. Policy makers should have as precise as possible conception of impacts that policy instrument will deliver, prior to its implementation. This is referred to as ***ex-ante*** or beforehand policy evaluation during which impacts (social, technological and financial) of policy instruments are forecasted. Expected impact in terms of reduced energy consumption and cost-effectiveness of the instruments are evaluated and compared to business-as-usual scenario in which no instruments are applied. However, often policymakers do not have enough experience and knowledge to confirm the established theory is right. Therefore, policymaking has to be publicly open process involving all stakeholders and market actors that could contribute to the overall understanding how the policy instrument is intended to work.

Unlike *ex-ante* evaluation of a policy, *ex-post* approach is applied after a certain time of the policy instrument implementation, effects of which should be evaluated to answer two key questions (Joosen& Harmelink, 2006):

- What was the contribution of policy instrument in the realisation of policy targets (**effectiveness of policy instruments**)?
  - Effectiveness of a policy instrument is measured as its net impact in the relation to the policy target set in the design phase. Net impact is equal to the difference between amount of energy used prior and after instrument is implemented. These are net energy savings but also related net CO_2 emission reductions that can be attributed to specific energy efficiency instrument taking free rider, spill over, rebound effect and other possible effects into account. Net impact is determined according to the previously defined baseline scenario.
- What was the **cost effectiveness of policy instruments**, and could targets have been reached against lower costs?
  - Cost effectiveness is the ratio between the additional costs caused by the instrument for the end-user, the society as whole or the government, and the net impact of the investigated instrument. Government costs are related to implementation, administration, enforcement of regulations, monitoring and evaluation, subsidies and tax relieves. In other words, cost effectiveness is used to determine how well public money is used to achieve socially beneficial goals. For end-users costs are determined by energy price, marginal investment and marginal operation and maintenance costs of energy efficiency measure.

However, instruments of energy efficiency policies might have other effects as well, so the third question it should be raised is:

- What **other impacts** did the policy achieved outside its main realm?
  - Most usually mentioned side effects of energy efficiency policy are environmental benefits and creation of new jobs, which are a positive effects in terms of ecological, social and economic stability and progress. However, sometimes negative effects are also possible to appear. E.g. CFLs are using far less energy and have longer life time and in a world's combat against climate change they are now starting to completely replace "old" incandescent light bulbs. However, CFLs do bring some other hazards, like small amount of highly toxic mercury they contain. Policy makers have to be aware of these relationships and often trade-offs have to be done - in this case, the trade-off has to be done between efficiency and potential health risk.

Answering these questions is referred as *ex-post* evaluation. It goes beyond evaluation of final delivered energy savings and tries to reveal success and failure factor enhancing in that way our knowledge about market performance. Enhanced knowledge gives the opportunity to improve effectiveness of policy instruments and to redefine our policy. Here both qualitative and quantitative assessments are needed and should be preferably supported by empirical data about policy performance. The backbone is cause-impact relationship, supplemented by indicators that measure the existence of cause-impact relationship, then failure and success factors should be listed (qualitative) and relationships with other policy instruments should be emphasised (other instruments can enhance or mitigate the impact of analysed instrument). In evaluation process empirical data are also very important as they are additional and often the only indicators of certain instruments impacts.

Both *ex-ante* and *ex-post* evaluation need to be supported with quantitative data, i.e. with data on energy efficiency improvements actually realised by implementation of policy instruments and energy efficiency improvement projects. The tools used for this purpose are referred to as **measurement and verification (M&V)** of energy savings. M&V is absolutely crucial part of any energy efficiency policy - it captures the overall improvement in energy efficiency and assesses the impact of individual measures. M&V procedures include two major methodological approaches: **top-down** and **bottom-up**. Both approaches must be combined to appropriately and as exact as possible evaluate the success of national energy efficiency policy and the magnitude of energy efficiency improvement measures' impact. Both approaches will be briefly explained hereafter, although it has to be emphasised that the detailed elaboration of M&V principles goes far beyond the scope of this chapter.

5.2. Top-down M&V methods

A top-down calculation method means that the amount of energy savings is calculated using the national or large-scale aggregated sectoral levels of energy saving as a starting point. This is purely statistical approach, often referred to as "energy efficiency indicators" because it gives an indication of developments.

Top-down methodology is based on collection of extensive data sets for not only energy consumption but also for various factors influencing it, and on calculation and monitoring of energy efficiency indicators. There are six types of indicators most commonly used. These are as follows[1].

1. **Energy intensity** - ratio between an energy consumption (measured in energy units: toe, Joule) and an indicator of activity measured in monetary units (Gross Domestic Product, value added). Energy intensities are the only indicators that can be used every time energy efficiency is assessed at a high level of aggregation, where it is not possible to characterize the activity with a technical or physical indicator, i.e. at the level of the whole economy or of a sector.
2. **Unit consumption or specific consumption** - relates energy consumption to an indicator of activity measured in physical terms (tons of steel, number of vehicle-km, etc.) or to a consumption unit (vehicle, dwelling …).
3. **Energy efficiency index (ODEX)** - provides an overall assessment of energy efficiency trends of a sector. They are calculated as a weighted average of detailed sub-sectoral indicators (by end-use, transport mode...). A decrease means an energy efficiency improvement. Such index is more relevant for grasping the reality of energy efficiency changes than energy intensities.
4. **Diffusion indicators** - there are three types of such indicators: (i) market penetration of renewables (number of solar water heaters, percentage of wood boilers for heating, etc.); (ii) market penetration of efficient technologies (number of efficient lamps sold, percentage of label A in new sales of electrical appliance, etc.); (iii) diffusion of energy efficient practices (percentage of passenger transport by public modes, by non motorised modes; percentage of transport of goods by rail, by combined rail-road transport, percentage of efficient process in industry, etc.). Diffusion indicators have been introduced to complement the existing energy

[1] These indicators are developed within ODYSSEE project and are used Europe- wide. More can be found at: http://www.odyssee-indicators.org/

efficiency indicators, as they are easier to monitor, often with a more rapid updating. They aim at improving the interpretation of trends observed on the energy efficiency indicators.

5. **Adjusted energy efficiency indicators** - account for differences existing among countries in the climate, in economic structures or in technologies. Comparisons of energy efficiency performance across countries are only meaningful if they are based on such indicators. External factors that might influence energy consumption include: (a) weather conditions, such as degree days; (b) occupancy levels; (c) opening hours for non-domestic buildings; (d) installed equipment intensity (plant throughput); product mix; (e) plant throughput, level of production, volume or added value, including changes in GDP level; (f) schedules for installation and vehicles; (g) relationship with other units. Some of these factors are relevant for correction of aggregated indicators, while some are to be used for the individual facilities in which energy efficiency measures are implemented.
6. **Target indicators** - aim at providing reference values to show possible target of energy efficiency improvements or energy efficiency potentials for a given country. They are somehow similar to benchmark value but defined at a macro level, which implies a careful interpretation of differences. The target is defined as the distance to the average of the 3 best countries; this distance shows what gain can be achieved.

The main advantages of the usage of top-down methods is their simplicity, lower costs and reliance on the existing systems of energy statistics needed for development of a country's energy balance. On the other hand, these indicators do not consider individual energy efficiency measures and their impact nor do they show cause and effect relationships between measures and their resulting energy savings. Developing such indicators requires huge amount of data (not only energy statistics, but whole set of macro and microeconomic data that are influencing energy consumption in all end-use sectors is needed), and data availability and reliability are often questionable in practice, sometimes leading to the huge need for modelling and expert judgement to overcome the lack of data. Nevertheless, energy efficiency indicators are inevitable part of energy efficiency evaluation process (both *ex-ante* and *ex-post*) as they are the only means to benchmark own performance against the performance of others, to reveal the potentials and help determine policy targets, to quantify the success/failure of the policy instruments and to track down the progress made in achieving the defined targets.

5.3. Bottom-up M&V methods

A bottom-up M&V method means that energy consumption reductions obtained through the implementation of a specific energy efficiency improvement measure are measured in kilowatt-hours (kWh), in Joules (J) or in kilogram oil equivalent (kgoe) and added to energy savings results from other specific energy efficiency improvement measures to obtain an overall impact. The bottom-up M&V methods are oriented towards evaluation of individual measures and are rarely used solely to perform evaluation of overall energy efficiency policy impacts. However, they should be used whenever possible to provide more details on performance of energy efficiency improvement measures. Bottom-up methods include mathematical models (formulas) that are specific for every measure, so only the principle of their definition will be briefly explained hereafter.

M&V approach boils down to the fact that the absence of energy use can be only determined by comparing measurements of energy use made before (baseline) and after (post-retrofit) implementation of energy efficiency measure or expressed in a simple equation:

$$\text{Energy Savings} = \text{Baseline Energy Use} - \text{Post-Retrofit Energy Use} \pm \text{Adjustments} \quad (2)$$

The baseline conditions can change after the energy efficiency measures are installed and the term "Adjustments" (can be positive or negative) in equation (2) aiming at bringing energy use in the two time periods (before and after) to the same set of conditions. Conditions commonly affecting energy use are weather, occupancy, plant throughput, and equipment operations required by these conditions. These factors must be taken into account and analysed after measure is undertaken and adjustments have to be made in order to ensure correct comparisons of the state pre- and post-retrofit. This kind of M&V scheme (often referred to as *ex-post*) may be very costly but they guarantee the detections of real savings. The costs are related to the actual measurement, i.e. to the measurement equipment. To avoid a large increase in the M&V costs, only the largest or unpredictable measures should be analysed through this methodology.
Individual energy efficiency projects might also be evaluated using well reasoned estimations of individual energy efficiency improvement measures impacts. This approach (*ex-ante*) means that certain type of energy efficiency measure is awarded with a certain amount of energy savings prior to its actual realisation. This approach has significantly lower costs and is especially appropriate for replicable measures, for which one can agree on a reasonable estimate. There are also some "hybrid" solutions that combine *ex-ante* and *ex-post* approaches in bottom-up M&V. This hybrid approach is often referred to as *parameterised ex-ante* method. It applies to measures for which energy savings are known but they may differ depending on a number of restricted factors (e.g. availability factor or number of working hours). The set up of a hybrid approach can be more accurate than a pure ex-ante methodology, without a substantial increase of the M&V costs.

5.4. Establishing evaluation procedures supported by M&V

The success of national energy efficiency policy has to be constantly monitored and its impact evaluated. Findings of evaluation process shall be used to redesign policies and enable their higher effectiveness. Regardless to its importance, policy evaluation is often highly neglected. Policy documents are often adopted by governments and parliaments and afterwards there is no interest for impacts they have produced. Therefore, setting up the fully operable system for evaluation of energy efficiency is a complex process, which requires structural and practice changes among main stakeholders in policy making. Additionally, it has to be supported by M&V procedures, which require comprehensive data collection and analysis systems to develop energy efficiency indicators that will quantify policy effects.

6. Conclusion

Evidently, energy efficiency policy making is not one-time job. It is a continuous, dynamic process that should create enabling conditions for energy efficiency market as complex system of supply-demand interactions undergoing evolutionary change and direct that change toward efficiency, environmental benefits and social well-being. However, there are number of barriers preventing optimal functioning of energy efficiency market, which should determine the choice of policy instruments. Policy instruments have to be flexible and able to respond (adapt) to the market requirements in order to achieve goals in the optimal manner, i.e. to the least cost for the society. Due to fast changing market conditions, Policy instruments can no longer be documents once produced and then intact for several years. Continuous policy evaluation process has to become a usual. Future research work to support policy making shall be exactly directed towards elaboration of methodology that will be able to qualitatively and quantitatively evaluate effectiveness and cost-effectiveness of policy instruments and enable selection of optimal policy instruments mix depending on current development stage of the energy efficiency market.

Evaluation procedures will advance and deepen our knowledge on success or failure factors of energy efficiency policy. The analysis of current situation shows that policies world-wide tend to fail in delivering desired targets in terms of energy consumption reduction. The main reason lies in the lack of understanding and focus on implementing adequate capacities, which are far too underdeveloped, insufficient and inappropriate for ambitious goals that have to be achieved. It has to be understood that policy implementation will not just happen by it self, and that capacities and capabilities in all society structures are needed. Embracing full-scale energy management systems in both public service and business sector can make the difference. Additionally, with the positive pressure from civil society organisations and media, understanding the interdependences of energy and climate change issues will improve, gradually changing the society's mindset towards higher efficiency, and eventually towards the change of lifestyle.

7. References

Morvaj, Z. & Bukarica, V. (2010). Immediate challenge of combating climate change: effective implementation of energy efficiency policies, *paper accepted for 21st World Energy Congress*, 12-16 September, Montreal, 2010

Morvaj, Z. & Gvozdenac, D.(2008). *Applied Industrial Energy and Environmental Management*, John Wiley and Sons - IEEE press, ISBN: 978-0-470-69742-9, UK

Dennis, K. (2006). The Compatibility of Economic Theory and Proactive Energy Efficiency Policy. *The Electricity Journal*, Vol. 19, Issue 7, (August/September 2006) 58-73, ISSN: 1040-6190

European Commission. (2006). *Action Plan for Energy Efficiency COM(2006)545 final*, Brussels

Eurostat. (2009). *Energy, transport and environment indicators*, Office for Official Publications of the European Communities, ISBN 978-92-79-09835-2, Luxembourg

European Environment Agency. (2009).*Annual European Community greenhouse gas inventory 1990–2007 and inventory report 2009*, Office for Official Publications of the European Communities, ISBN 978-92-9167-980-5, Copenhagen

European Commission. (2009). *Draft Communication from the Commission to the Council and the European Parliament: 7 Measures for 2 Million New EU Jobs: Low Carbon Eco Efficient & Cleaner Economy for European Citizens*, Brussels

Bukarica, V.; Morvaj, Z. & Tomšić, Ž. (2007). Evaluation of Energy Efficiency Policy Instruments Effectiveness - Case Study Croatia, *Proceedings of IASTED International conference "Power and Energy Systems 2007"*, ISBN: 978-0-88986-689-8, Palma de Mallorca, August, 2007, The International Association of Science and Technology for Development

Briner, S. & Martinot, E. (2005). Promoting energy-efficient products: GEF experience and lessons for market transformation in developing countries. *Energy Policy*, 33 (2005) 1765-1779, ISSN: 0301-4215

Vine, E. (2008). Strategies and policies for improving energy efficiency programs: Closing the loop between evaluation and implementation. *Energy Policy*, 36 (2008) 3872–3881, ISSN: 0301-4215

Bulmstein, C.; Goldstone, S. & Lutzenhiser, L. (2000). A theory-based approach to market transformation, *Energy Policy*, 28 (2000) 137-144, ISSN: 0301-4215

Paskaleva, K. (2009). Enabling the smart city: The progress of e-city governance in Europe. *International Journal of Innovation and Regional Development*, 1 (January 2009) 405–422(18), ISSN 1753-0660

Stanislaw, J.A. (2008). *Climate Changes Everything: The Dawn of the Green Economy*, Delloite Development LCC, USA

Morvaj, Z. et al. (2008). Energy management in cities: learning through change, *Proceedings of 11th EURA conference, Learning Cities in a Knowledge based Societies*, 9-11 October 2008, Milan

Joosen, S. & Harmelink, M. (2006). *Guidelines for the ex-post evaluation of 20 energy efficiency instruments applied across Europe*, publication published within AID-EE project supported by Intelligent Energy Europe programme.

2

Energy growth, complexity and efficiency

Franco Ruzzenenti* and Riccardo Basosi*°
**Center for the Studies of Complex Systems, University of Siena*
°Department of Chemistry, University of Siena
Italy

1. Introduction

Over the last two centuries, the human capacity to harness energy or transform heat into work, has dramatically improved. Since the first steam engine appeared in Great Britain, the first order thermodynamic efficiency (the rate of useful work over the heat released by the energy source) has soared from a mere 1 % to the 40 % of present engines, up to the 70% of the most recent power plants. Despite this efficiency revolution, energy consumption per capita has always increased (Banks, 2007).

The economy and society have undeniably faced an expanding frontier, and both household and global energy intensities have commonly been linked to economic growth and social progress. The rising issue of energy conservation has prompted us to consider energy efficiency as more than merely a characteristic of economic growth, but also as a cause (Ayres and Warr, 2004). We thus wonder if it is possible to increase efficiency, reduce global energy consumption, and foster economic development within an energy decreasing pattern, by separating efficiency and energy growth. In other words, by reducing efficiency positive feed-backs on the system's energy level (Alcott, 2008).

In 1865, the economist Stanley Jevons was the first to point out the existence of a circular causal process linking energy efficiency, energy use, and the economic system. Jevons was convinced that efficiency was a driving force of energy growth and highlighted the risk associated with an energy conservation policy thoroughly committed to efficiency[1].

Recently the Jevon's paradox has been approached in the field of Economics and termed "rebound effect". It has been the subject of articles, research, as well as a great deal of controversy over the last two decades (Schipper, 2000). Although many economists are still sceptical as to its actual relevance, most of them have agreed on the existence and importance of such an effect. Some are deeply concerned (Khazzoum, 1980, Brookes 1990,

---

1 "It is very commonly urged, that the failing supply of coal will be met by new modes of using it efficiently and economically. The amount of useful work got out of coal may be made to increase manifold, while the amount of coal consumed is stationary or diminishing. We have thus, it is supposed, the means of completely neutralizing the eveils of scarce and costly fuel. But the economy of coal in manufacturing is a different matter. *It is a wholly confusion of ideas to suppose that the economical use of fuel is equivalent to a diminished consumption. The very contrary is the truth* (Jevons, 1965)."

Saunders, 2000, Herring, 2006) about the overall net effect and its capacity to counterbalance the gains due to efficiency. Others, however, still believe in the net benefit of energy policies focused on developing energy efficiency, although they admit the burden of having to pay a loss of savings (Shipper and Haas, 1998; Washida, 2004; Grepperud & Rasmussen, 2003).
The most accurate and simple definition of rebound effect is: a measure of the difference between projected and actual savings due to increased efficiency (Sorrell and Dimitropoulos, 2007).
Three different kinds of rebound effects are now widely used and accepted(Greening and Greene,1997):

1. Direct effects: those directly linked to consumer behaviour in response to the more advantageous cost of the service provided. They depend on changes in the final energy use of appliances, devices or vehicles (i.e. if my car is more efficient, I drive longer).
2. Indirect effects: those related to shifts in purchasing choices of customers, either dependent on income effects or substitution effects, which have an ultimate impact on other energy services (i.e. new generation engines are economical, then I buy a bigger car or I spend the money saved for an air conditioner).
3. General equilibrium effects: changes in market demands as well as in relative costs of productive inputs that ultimately have a deep impact in the productive structure, possibly affecting the employment of energy as a productive factor (i.e. the well known substitution of capital to labour, subsequent to a rise of labour costs, is otherwise an increase of the energy intensity of the system. Labour cost may increases relative to a subsidiary process that employs more energy to run).

The above classification displays the circular feedback process's (increasing) time lag scheme, beginning with a quick response, the altered use of energy devices due to changes in energy costs, followed by a slower mechanism, changes in purchasing choices, and finally, the long term restructuring process affecting economic factors. While direct and indirect effects have found considerable attention in the literature, general equilibrium effects remain relatively unexplored due to the uneasiness of their time scale and the variety of involved variables (Binswanger, 2001)[2].

2. The economic approach to the rebound effect

However paradoxical the rebound effect may seem, it can be explained by classic economic theory. Energy is a derived demand because it is not the actual good purchased, but a means by which a good or a service is enjoyed. Thus, technology that is able to reduce the amount of energy employed by good or service lowers the cost of that item. It is said that efficiency improvements reduce the implicit price of energy services and, according to the basic theory of market demands, the amount of goods consumed rises when prices decrease. Happy with this explanation, economic theory focused on measuring and forecasting the rebound effect. Both econometric models and neoclassical forcasting models have been

---

2 "Third, changes in the prices of firms' outputs and changes in the demand for inputs caused by income and substitution effects will propagate throughout the economy and result in adjustments of supply and demand in all sectors, resulting in general equilibrium effects. By taking care of the income effect, we also include the indirect rebound effect in our analysis, but we still neglect general equilibrium effects (Binswanger, 2001).

developed that exhibit sound results, except for the third kind of effect, that unfortunately presents many features unfit for these models (Saunders, 1992; Greening and Greene, 1997; Binswanger, 2001; Sorrell, 2009).

Forecasting models are mainly based on Cobb-Douglas production functions, with three factors of production (capital, labour, energy), and which derive market demands for these factors. Since the first attempts, calculations confirmed the existence of the effect under the assumption of constant energy prices (Saunders, 1992). Econometric based research also verified the relevance of the rebound effect and further provided valid measures of the effect in a variety of economic sectors. Such measures mainly utilize the relative elasticities of demand curves. Demand curves are built on statistical regressions in prices and quantities of goods, while elasticity is a measure of the sensitivity in demand to the variation of a good's price. Although these models may be accurate, they are all single good or service designed and are consequently viable only for the detection of direct effects.

Other models based on substitution elasticities between goods or factors as well as income elasticities have addressed indirect effects (Greening and Greene, 1997). Such contributions brought the level of detection to a whole sector of an economy or to a variety of aspects related to the process of substitution highlighted in the rebound effect like the role of time-saving technologies and their impact on energy intensities (Bentzen, 2004; Binswanger, 2001). Nevertheless, very few attempts have been made to evaluate general equilibrium effects, a task which entails the recognition of the main connecting variables of an economy, spread over a long period of time. These contributions, however, fail to describe and explain major structural changes in the productive systems that cause discontinuity in the economic relations among variables. All these models are, in fact, based on a stationary framework, and therefore neglect evolutionary changes that heighten the developing pattern of an economy (Dimitropoulos, 2007).

As a result of being the first who introduced the paradox behind the development of efficiency, Jevons' work has to be considered a landmark in this matter, for he was able to trace a line that goes beyond the mere economical, or the implicit price mechanism, explanation. He thought that any technological improvement rendering the energy source more economical would stimulate the demand for energy. Furthermore Jevons had some advanced and valuable intuitions about the role of energy sources in the economic development, as well as about the dynamic between technology, energy and the economy that were too often neglected by modern economists. His contributions are summarized as follows:

1. Fuel efficiency affects market size and shape, and not just a process of substitution among factors. He noticed that both time scale and space scale of travels changed with engine technologies making new markets or new places reachable[3].
2. Features of energy sources other than efficiency are relevant for economic purposes like energy intensity and time disposal (power). He argued that what made steam

[3] Such structural changes are unfit for common, wide spread modeling approaches. Is noteworthy that when Jevons was developing his analysis, consumer theory was far to come and main sectors were those of steal, mining and machinery industries. Economy was chiefly engaged in building his back bone and changes at any rate were basically *structural*. His view of economic processes was consequentially affected by that turmoil and can be considered, to a certain extent, *evolutionary*. Shipper has raised the attention on structural changes, which are, according to his opinion, hardly detectable but very important in energy demand long term pattern (Shipper and Grubb, 2000).

vessels more economical was neither fuel efficiency (wind power is more efficient) nor unit costs (wind vessels are almost costless), but instead the availability and disposal of coal as an energy source which had an incomparable positive impact on the capital return cycle.

3. A sink or a flux of free energy becomes an energy source when there is an exploiting technology and an economic need forward. He argues that from the beginning onward, a developing process of energy sources has a fundamental role as an economic driving force and not vice versa. In other words, when economic needs are compelling, technology development is significantly accelerated and as a result, feeds back to the whole economic system.
4. Prosperity is dependent on economical energy sources, and economic development is mainly shaped by energy sources and its quantity[4]. However pessimistic we may consider this statement, Jevons meant to call for an economical austerity in order to prevent society form a hard landing due to the running out of low cost coal[5]. He claimed that was more recommendable a stationary economy together with social progress.

What we can therefore gain from his teachings is that there is an inner tendency of an economy to render energy sources more economical and that this is the true driving force of economic development[6].

Thus, for Jevons, societal development—civilization—is "the economy of power" or the constant strain on humanity of harnessing energy in a productive way, and its "history is a history of successive steps of economy (energy efficiency, n.d.r.)." The incremental process

---

4 "We may observe, in the first place, that almost all the arts practiced in England before the middle of the eighteenth century were of continental origin. England, until lately, was young and inferior in the arts. Secondly, we may observe that by far the grater part of arts and inventions we have of late contributed, spring from our command of coal, or at any rate depend upon its profuse consumption" (Jevons, 1965).

5 A misleading, wide spread, opinion is that Jevons skepticism was misjudged and the rising age of oil gave proof of it; but he clearly foresaw the drawbacks of such a solution: "Petroleum has, of late years, become the matter of a most extensive trade, and has been found admirably adapted for use in marine steam-engine boilers. It is undoubtedly superior to coal for many purposes, and is capable of replacing it. But then, *What is Petroleum but Essence of Coal,* distilled from it by terrestrial or artificial heat? Its natural supply is far more limited and uncertain than of coal, and an artificial supply can only be had by the distillation of some kind of coal at considerable cost. To extend the use of petroleum, then, is only a new way of pushing the consumption of coal. It is more likely to be an aggravation of the drain then a remedy."

6 "The steam-engine is the motive power of this country, and its history is a history of successive steps of economy. But every such improvement of the engine, when effected, does but accelerate anew the consumption of coal. Every branch of manufacture receives a fresh impulse-hand labour is still further replaced by mechanical labour, and greatly extended works can be undertaken which were not commercially possible by the use of the more costly steam-power. But no one must suppose that coal thus saved is spared –it is only saved from one use to be employed in others, and the profits gained soon lead to extended employment in many new forms. The several branches of industry are closely interdependent, and the progress of any one leads to the progress of nearly all. And if economy in the past has been the main source of our progress and growing consumption of coal, the same effect will follow from the same cause in the future."

of energy efficiency drives more and more energy into the system, but how does it occur? Jevons, in the following passage, provides insight into such a controversial question:
Again, the quantity consumed by each individual is a composite quantity, increased either by multiplying the scale of former applications of coal, or finding new applications. We cannot, indeed, always be doubling the length of our railways, the magnitude of our ships, and bridges, and factories. In every kind of enterprise we shall no doubt meet a natural limit of convenience, or commercial practicability, as we do in the cultivation of land. I do not mean a fixed and impassible limit, but as it were an elastic limit, which we may push against a little further, but ever with increasing difficulty. But the new applications of coal are of an unlimited character (Jevons, 1965).

3. Complexity and Efficiency

Jevons believed that the natural tendency of economy is to expand linearly, "multiplying the scale of former applications," up to a limit and then, to overcome such limits, the system works within itself to develop "new applications". Sketched roughly, the scheme here is: growth-saturation-innovation-growth.
Jevons found an unsuspected counterpart in a famous biologist, Alfred Lotka, who was interested in the relation between energy and evolution. Indeed there are several analogies among their theories. Lotka too believed in the need for looking synoptically at the biological system in order to understand the energetics of evolution. Lotka also shares Jevons' cyclic view of processes, which, in the case of energy "transformers," he understood to be formed by an alternation growth-limit to growth- evolution- growth[7]. According to Lotka, the reason why this process was doomed to an ever growing amount of energy flow boiled down to the cross action of selection-evolution on the one hand and the thermodynamics law on the other. In his opinion, evolution is the result of a stochastic process and a selective pressure, and moreover, "the life contest is primarily competition for available free energy." Thus, selection rewards those species adapted to thrive on a particular substrate, and the growth of such species will divert an increasing quantity of free energy into the biological system. Those species' growth will proceed until the free energy available for that transformation process is completely exploited. The dual action of case and selection will then favor new transformers more efficient in employing the free energy still available. The developmental stages of ecological succession mirror this evolutionary energetic pattern. In the first stage of ecological succession, plant pioneering species dominate, growing rapidly, but inefficiently disposing of resources. In the climax stage,

---

7 "But in detail the engine is infinitely complex, and the main cycle contains within its self a maze of subsidiary cycles. And, since the parts of the engine are all interrelated, it may happen that the output of the great wheel is limited, or at least hampered, by the performance of one or more of the wheels within the wheel. For it must be remembered that the output of each transformer is determined both by its mass and by its rate of revolution. Hence if the working substance, or any ingredient of the working substance of any of the subsidiary transformers, reaches its limits, a limit may at the same time be set for the performance of the great transformer as a whole. Conversely, if any one of the subsidiary transformers develops new activity, either by acquiring new resources of working substance, or by accelerating its rate of revolution, the output of the entire system may be reflexly stimulated

however, the most efficient species in converting resources prevail (Odum, 1997). The following passage stresses this key concept:
This at least seems probable, that so long as there is abundant surplus of available energy running "to waste" over the sides of the mill wheel, so to speak, so long will a marked advantage be gained by any species that may develop talents to utilize this "lost portion of the stream". Such a species will therefore, other things equal, tend to grow in extent (numbers) and its growth will further increase the flux of energy through the system. It is to be observed that in this argument the principle of the survival of the fittest yields us information beyond that attainable by the reasoning of thermodynamics. As to the other aspect of the matter, the problem of economy in husbanding resources will not rise to its full importance until the available resources are more completely tapped than they are today. Every indication is that man will learn to utilize some of the sunlight that now goes to waste (Lotka, 1956).
Economy and biology are both evolutionary systems and both can be approached from thermodynamics. By contrast, not all analogies are suitable. Whilst *less efficient* transformers like bacteria persist together with more evolved vertebrates, hence biosphere makes manifest the entire evolutionary path, economy dismisses obsolete technologies (we don't see any more steam motive engines around). So, if we abandon inefficient technologies, why isn't the net effect over consumptions negative? In other words why, if we employ more efficient devices, energy use doesn't drop? History has so far proved that more efficiency results in more energy consumption. Where does this paradox come from?
Is this paradox due to the counteractive effect of population or affluence growth over efficiency or is efficiency evolution the driving factor of economic growth? We will here attempt to show how the causality chain initiate with an efficiency improvement and that growth comes after. Growth featured by those changes affecting the economic system comparable to "new applications of unlimited character" mentioned by Jevons or an "acceleration to the revolution rate of the world engine" envisioned by Lotka.
What it is being argued here is that all those changes, or among them, those affecting the structure or delivering brand new technologies into the system, may be regarded as a leap of complexity occurring to the system. Complexity, in the acceptation of organizational complexity, if it was observed as a feature of whatsoever of a system, has always displayed a high energy density rate. This means that growing complexity implies growing energy consumption. That is to say, a more complex system consumes more (more connections, more variety, more hierarchical levels). It is therefore possible that the energy saved by new and more efficient processes is absorbed or perhaps a better word, dissipated, by a more complex system. Energy savings resulting from increased efficiency would then be offset by an organization restructuring process within the system.

4. Evolutionary Pattern

We have advanced the hypothesis of the existence of a common, recursive pattern in evolutionary systems. This pattern underlies a broad, complex thermodynamic process involving the entire system and arises from forces embedded within the system. We have described this pattern as the following circular process: **growth-saturation-complexity leap-growth** and can be depicted it as a circular process.

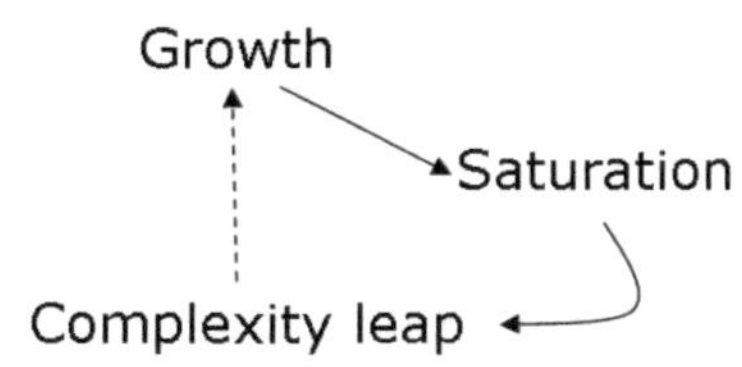

Fig. 1. Evolutionary Pattern

The growth stage relies on the presence of inner forces that drive the system to expand while seeking survival and reproduction. These forces are species (the genome) in the domain of biology, and firms (the capital) in the economy. Although it is clear how these autocatalytic processes cause the system's expansion, it is less clear how, coupled with efficiency improvements, they can divert more energy into the system or in the words of Lotka, "maximize the energy flow." It must be kept in mind that neither Lotka nor Jevons claims that the overflow of energy is the actual aim of system components. It is rather a result of their interaction with each other and with the environment. Lotka, for example, believes that two main thermodynamic strategies are adopted by organisms in order to adapt to the environment: maximizing output (power maximum) and minimizing input (efficiency maximum). The former is developed by species thriving in resource abundance and the latter by organisms struggling in scarcity conditions. According to Lotka, by pursuing unexploited free-energy more energy is driven through the system thus maximizing global output. The dichotomy between efficiency and power is therefore quite apparent[8].

And there is indeed something well founded in this revelation, which is rooted in thermodynamics. The antagonism between efficiency and power is less evident from a thermodynamics perspective, meaning that if other factors are left unchanged, an efficiency improvement always leads to empowerment. The misunderstanding and thereby the paradox of efficiency comes from two major misconceptions, which can be outlined as follows:

- Thermodynamic efficiency, from the Carnot Engine onward, concerns the conversion of heat into work, not just the mere transformation of one form of energy to another.
- Efficiency, as a rate between output and input or benefits and costs, pertains to a static analysis despite the fact that the conversion process actually takes place in time and therefore costs and benefits also depend on the time elapsed.

[8] There is a simplification of Lotka's vision of the energetics of evolution that states that two strategies would top evolutionary thermodynamics: one that maximizes work over time (power) in the case of resource abundance and another that minimizes energy consumed per for amount of work delivered (efficiency) in the case of scarcity. These two strategies have been summarized in the "maximum power principle," despite Lotka himself being reluctant to adopt any lofty and ambitious term like "principle" for his thinking. Moreover, in this formulation, scarcity and abundance are unrelated whatsoever to magnitude, while Lotka clearly stresses what scarcity must be compared to: the ability of a transformer to get hold of free energy and its growing rate. What are indisputably scarce or plenty are nutrients, row materials or water, which eventually affect energy efficiency.

The first statement assumes the custom of considering conversion rates, such as the transformation of chemical energy into heat, as thermodynamic efficiencies. As previously noted, most of the controversies surrounding the rebound effect in the residential sector arise from the misleading concept of efficiency. The rate of transformation of chemical energy into heat in e.g. a bomb calorimeter is a calorie while out of the laboratory, it is a thermal efficiency, and should not be considered a thermodynamic efficiency because no work is involved[9]. The theoretical apparatus we have so far employed is therefore inapplicable. Only work needs an entropy change into the (work) reservoir in order to be dissipated while a heat sink is of unlimited disposal to the environment. In other words, the system's structure needs to change in order to dissipate (more) mechanical work, but not the same can be said for heat. This kind of efficiency, known as thermal efficiency, has much more to do with squandering. When a process becomes more thermodynamically efficient, more work is extracted from the same amount of energy (heat) and when it becomes more thermally efficient, less heat for our purpose is wasted from a heat source.

4.1 The Time Variable Determines the Efficiency Level

In the second statement, the attention is focused on a theoretical aspect that needs a formal treatment to be understood. It is indeed very difficult to intuitively sense that, in physics terms, a system that improves its efficiency also enhances its power. It is even more difficult to see how this can be true if a trade off exists between power maximization and efficiency optimization. A system that maximizes its efficiency actually minimizes its power and vice versa. Thus, if we improve the efficiency, we increase the power. Nevertheless, if we seek the best efficiency, we have to set the minimum power output. Is this a paradox? In a sense, yes, but only if our analysis is oblivious to the passage of time.

We have formulated two assertions in apparent contradiction. The first is that when thermodynamic efficiency improves, power increases. This direct relationship is evident by observing the definitions of efficiency and power:

$$\eta = \frac{W}{Q_h}, P = \frac{W}{\Delta t} \tag{1}$$

As long as the specific consumption—the rate at which the energy source is depleted—remains constant, the power increases. It is noteworthy that this relationship strictly relates to the capacity of the system to draw from a particular source. The capacity depends on the specific consumption:

$$\frac{\partial Q_h}{\partial t} \tag{2}$$

The specific consumption is the rate of depletion of the energy source or the amount of input (fuel) per the unit of time. It reflects the capacity of the system to convey energy

9 Thermodynamic efficiency concerns the transformation of heat into work. Other *non-thermodynamic* efficiencies are, for example, heat transport and heat regulation or the cinematic chain. Nevertheless, any *kind of efficiency* can contribute to the overall thermodynamic efficiency, when a work output is obtained out of heat.

throughout the process.The second assertion that there exists a trade-off between *efficiency* and *power* needs more mathematics to be explained. It will be illustrated by means of a Carnot Cycle, revisited with the addition of the time variable. In the Carnot Cycle, to achieve the maximum efficiency, the isothermal expansion and compression (Figure 2), need to occur at an infinitely slow speed in order to maintain an infinitesimal temperature gradient between the working substance (T_{hw}, T_{cw}) and the heat reservoirs (T_h, T_c). Under these circumstances, the power of the machine approaches zero since it takes infinite time to produce a finite amount of work. To speed up the process, we need to increase the gradient since the heat transfer rate is proportional to it. To thereby get more than an infinitesimal amount of power from a Carnot Engine, we have to keep the temperature of its working substance below that of the hot reservoir and above that of the cold reservoir.

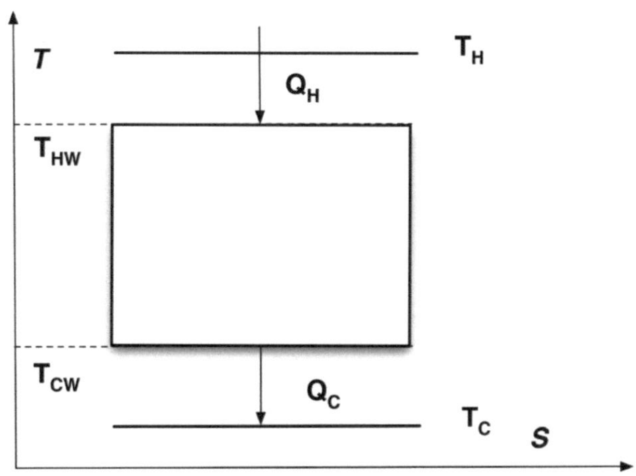

Fig. 2. Carnot Cycle

The more we increase the two gradients, the closer the extreme temperatures of the working substance. Ultimately, the two isothermal stages take place with no change in the temperature of the working substance. Heat flows directly from the hot source to the cold sink and no work is done. Hence the power output is zero and the engine has zero efficiency as well. In this model, we consider a Carnot Engine with a working substance absorbing heat from the hot source at T_{hw} and releasing heat to the cold source at T_{cw}. Under most circumstances, the rates of heat transfer will be proportional to the temperature gradients. We assume the constant of proportionality (K -meaning that heat absorption/release occurs in the same conditions) and the same Δt for the expansion and the compression[10]. We also assume that the two adiabatic transformations remain unaltered. We now have the following equations describing the *once* isothermal processes:

$$\frac{Q_h}{\Delta t_1} = k\left(T_h - T_{hw}\right) \tag{3}$$

[10] These assumptions can be abandoned without changing the results of the model, see Curzon and Ahlborn (Curzon and Ahlborn, 1975).

$$\frac{Q_c}{\Delta t_2} = k(T_{cw} - T_c) \tag{4}$$

T_h= temperature of the hot source, T_c=temperature of the cold source, T_{hw}=max temperature of the working fluid, T_{cw}=min temperature of the working fluid
Since the remaining two processes are adiabatic, they follow the relation (5):

$$\frac{Q_h}{T_{hw}} = \frac{Q_c}{T_{cw}} \tag{5}$$

The power of the system will be defined in equation (6):

$$P = \frac{W}{2\Delta t} \tag{6}$$

$$W = Q_h - Q_c \, , \quad \Delta t_1 = \Delta t_2 \tag{7}$$

The maximization of the power, as a function of T_{hw}, the hotter working temperature, will give the following result for the optimum power output:

$$T^{hw} = \frac{1}{2}\left(T_h + \sqrt{T_h T_{cw}}\right) \tag{8}$$

at a corresponding efficiency of

$$\eta = 1 - \sqrt{\frac{T_c}{T_h}} \tag{9}$$

It will be useful to do a variables' substitution to depict the trade off so we now fix $x = T_{cw}/T_{hw}$. According to this model, the efficiency-power trade off can be sketched as function of x and whereby Carnot efficiency will be represented by curve (10) and power output curve (11):

$$\eta = 1 - x \tag{10}$$

$$P = \frac{k}{4}\left(T_c + T_h - T_h x - \frac{T_c}{x}\right) \tag{11}$$

The two curves can be plot in a graph, assuming T_h and T_c of 300 and 25 degree Celsius; and fixing k at 0.05 (Fig.3). To reach the maximum theoretical efficiency (η for the isothermal transformation) the system must approach thermal equilibrium and therefore maximum slowness. Since it arises from power maximization, the optimal output will be somewhere between theoretical maximum efficiency and zero efficiency and it will only be determined by the sources' temperatures (T_h and T_c). So for every boundary condition in a Carnot Cycle, there is a single optimal value of output. Even if we abandon most of the abstract assumptions about the Carnot Cycle thus introducing further irreversibility, the peak of the curve will probably shift, but the trade off is unavoidable. We have to set the engine at either maximum efficiency or maximum power. "However, when the cost of building an engine is much greater than the cost of fuel (as is often the case), it is desirable to optimize the engine for maximum power output, not maximum efficiency (Schroeder, 2000)."

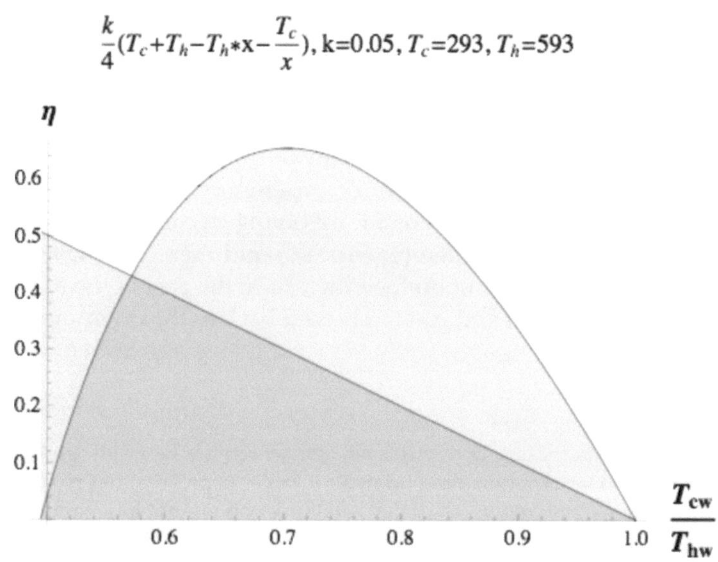

Fig. 3. Power-efficiency trade off

The power maximization will lead to sub-optimal efficiency (with respect to Carnot efficiency) which depends on sources' temperatures with the explicit relation (9) while Carnot efficiency is:

$$\eta_{Carnot} = 1 - \frac{T_c}{T_h} \tag{12}$$

It is noteworthy that such an efficiency level seems to be much closer to the running efficiency of most of energy converting sources than the Carnot efficiency (Table 1).

4.2 Efficiency improvement and power enhancement

We can further assume that efficiency improvements also apply to engine parts, in addition to working temperatures[11]. Any technical improvement concerning the material employed

[11] If we consider sources' temperature changes, we return to the dominion of Carnot efficiency while if we take into account working temperatures, we resort to the efficiency-power trade off sketched by the model.

or the reduction of friction would lead to a higher K and a better (faster) heat transfer across the machinery[12]. Since K does not affect the output regulation (the maximum value is not dependent on K), this will in turn, increase the rate of Q_h and the power. According to the value of the maximum power, it is clear that any increase in K (given T_h and T_c) will augment the power by a factor of 1/4, shifting the peak upward. More efficiency will therefore lead to higher power.

Suppose we want to increase efficiency as much as possible, leading the control parameter K. We may push further K, in order to increase the heat transfer rate and get an higher efficiency, but we will end up moving away from the theoretical maximum efficiency level, toward an higher power output, as it is shown in the animation[13].

Through this model, we have shown how the efficiency-power paradox is apparent and we have also described the thermodynamic conditions of the efficiency-power trade off. We can thus draw Lotka's conceptual framework of power maximization versus efficiency optimization in the context of the economy of power. It is, as we have already highlighted, an *economic* optimization that leads to maximum output[14]. Whenever the cost of fuel is relatively less constraining than the cost of machinery, power will be maximized. Nevertheless, every efficiency improvement involving technological development will probably lead to a more complex engine (or process) and therefore, will, on the one hand, reduce the relative price of energy, but on the other, raise the cost of the apparatus. This will ultimately amplify the bifurcation and positively feed back to the optimum power level[15].

---

[12] The paradoxical effect of increasing both efficiency *and power* can be easily understood if we think energy as space integral and work as the time integral of force. A process that reduces energy input in less time, increases power, as integrations over the same function are not independent. That is to say: if we use less energy per unit of space, and unit of time, in the same amount space we will save energy, yet in the same *time lag* we may use more energy!

[13] Animation at: http//sciyo.com

[14] Concepts of the like of "costs" and "economic optimization" should not be intended in a strict way. Broadly speaking, costs are to be meant as thermodynamic cost.

[15] The idea of sub-optimal efficiency level output was investigated in the filed of biological systems. As early as the 1955, Hodum and Pinkerton (Hodum H.T., Pinkerton R., 1955) published an article in which, adopting Lotka approach an vision for life's energetics, tried to demonstrate that "natural systems tend to operate at that efficiency which produces a maximum power output". Such efficiency was lower then the maximum attainable and, according to them, was exactly of 50%. "In natural systems there is a general tendency to sacrifice efficiency for power output". The idea of the 50% set point was based on the finding that most of energy converting systems were featured by coupled antagonists processes. "The essence of biochemical workings of an organism is the coupling of an exergonic catabolism to an endergonic anabolism that results in growth, reproduction and maintenance". Although this paradigm may account partially or even totally, for the derivation of the 50% value of efficiency, it was a striking intuition. It is remarkable, for example, that in the former model, whereas it is not so evident, there are two counteractive processes: the heat absorption an then heat release. The heat disposal affect the power output as much as the heat intake, as we know empirically, from electric power plants. Thus, this simple thermodynamic model resembles, by this point of view, the "living systems" of Odum theory. Conversely, as already Odum didn't fail to mention later on (Odum, 1983), the article of Cuzon and Ahlborn of 1975, on which this model is based, gave a sound evidence and a formal basis, to the postulate of the "maximum power principle"

Power Source	Tc	Th	η (Carnot)	η* (model)	η (observed)
Coal Fired Steam Plant	25	565	64.10%	40%	36.00%
Nuclear Reactor	25	300	48.00%	28%	30%
Geothermal Steam Plant	80	250	32.30%	17,5%	16%

Table 1 Source: Cuzon and Ahlborn, 1975.

4.3 The Case of Trucks

The truck industry and therefore, the road freight transport sector, gives a useful example of empowerment brought about by the efficiency improvement of the engine and vehicle technologies. From the late 1970's onward, efficiency rose steadily as an effect of technology research that tried to overcome the effects of soaring energy costs. Initially, such improvements were employed to reduce consumptions, but later technology development partially addressed power enhancement. Energy efficiency, as measured by fuel economy - distance travelled, at constant speed, for unit of fuel consumed, increased since late 1970's to late 1990's of about 30%. However, if we rescale fuel economy to the power shift of engines (adjusted fuel economy), we can observe a major change in efficiency (Ruzzenenti and Basosi, 2009a). This is also evident from the comparison of two trends of fuel economy and adjusted fuel economy (fuel economy divided by the engines' power) for a sample of 97 different European heavy-duty trucks. Initially the two metrics are coupled and show how efficiency was employed to reduce fuel consumption; we can see a dramatic drop in both fuel economy and adjusted fuel economy. Later trends display a sharp bifurcation, from mid 1990's onward, that explains how efficiency was then employed to enhance power and reduce consumptions (Figure 3)

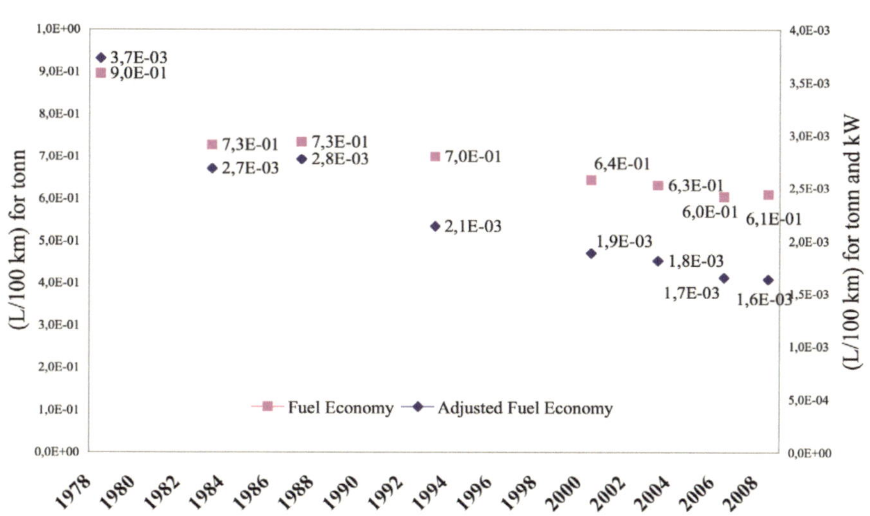

Fig. 4. Efficiency and power bifurcation in European Truck Industry (Ruzzenenti and Basosi, 2009a)

5 Structural Complexity

The underlying hypothesis of this work is that higher complexity counterbalances, on a global scale, the effects of higher efficiency on a process scale. It is our general understanding of evolution that selection operates by reward complexity. More complex, in the context of biology is often used as a synonym for fittest in terms of the competition for resources. Technological advances also develop from less to more complex devices. The meaning of complexity has never been questioned, for it has been evident in the semantic of nature or progress since earliest observations. A eukaryotic cell is more complex than a prokaryotic one and a Ferrari F1 is more complex than a Ford T. Under this perspective, complexity is countable, if not measurable, by the number of different components, parts or organs. If we abandon the conviction that progress always evolves toward higher complexity, that is to say, if we relinquish the belief of an immanent trend in nature; or if we are dealing with systems that differ in structure rather than in number of components, how can we apply such well established knowledge?

It is beyond the goals of this analysis to establish what complexity is or how it should be approached. The scientific community has been unable to establish or agree on a universal definition or paradigm of complexity and any attempt to univocally measure complexity is therefore doomed to failure[16]. It will be here assumed that a more complex system has a higher energy density or in other words, *consumes more energy per unit of mass and time*. There is a great deal of evidence, from biological records to cosmological entities, of such a relationship and therefore we believe it can be considered a *reasonable assumption*.

Indeed, this strongly recursive pattern in nature –linking energy density and compolexity, caused many scientists to think that energy could itself be considered a measure of complexity (Odum H.T., 1996; Odum E.P., 1997; Chaisson, 2001). Let's assume a more complex system consumes more (per unit of mass and time) and the complexity we refer to, it is *a structural* or *morphological* complexity, as we are dealing with systems with undefined boundaries and innumerable components, like the productive and transport systems. The two main assumptions regarding complexity that we are concerned with are:

1. A more complex system consumes more energy per unit of mass and unit of time (higher energy density rate)
2. Structural complexity primarily concerns the components' organization[17] rather than the components' variety or number

---

[16] "By 'complexity', we refer to the term intuitively as used in ordinary discourse, a definition culled from many sources: 'a state of intricacy, complication, variety, or involvement, as in the interconnected parts of a structure-a quality of having many interacting, different components.'In this work we shall come to identify complexity in two operational ways: as a measure of the information needed to describe a system's structure and function, or as a measure of the rate of energy flowing through a system of given mass. No attempt is made here to be rigorous with the words 'order', 'organization', 'complexity', and the like; this is not a work of classical philology or linguistic gymnastics. Indeed, no two researchers seem able to agree on a precise, technical definition of such a specious word as complexity, which may be context-dependent in any case (Chaisson, 2001)".

[17] For organization we refer to any system's components acting or arranged in a cooperative, systematic fashion.

Further remarks attain the duality efficiency/complexity. We should bear in mind that while we are referring to energy efficiency improvements, we are dealing with a process-scale analysis, whilst the leap in complexity concerns the global-scale analysis. These phenomena are at two different hierarchical levels:

1. Energy efficiency concerns energy converting processes and is therefore at the components level of the system
2. Complexity (structural) concerns the organization of the system and is thus at the global level of the system

We try to hereafter relate energy efficiency enhancement to complexity change. To accomplish this, we have to detect changes in system organization that move in the direction of higher complexity. Yet, changes in *which* system?

Since we have been dealing in the case study with truck efficiency, it would make sense to refer to the *freight road transport system,* but that would be misleading because the goods transport sector is merely a sub-component of the whole productive system. Transport service is just a *derivative* demand, which means, in economics terminology, that someone wants a good to be moved from one place to another. The shipment is the means, not the goal. Our analysis therefore has to address the productive system in order to detect long term changes in transport demand. Transport demand is *derived* from the needs of the productive system. The transport system and the productive system, under the scope of present analysis, are two parts of a whole.

5.1 Complexity leap: structural analysis

The main feature of the shift from a *fordian* to a *post-fordian* system concerns the location of the productive chain. Formerly, the productive chain was set entirely in one site, to which raw materials were delivered and from which products were shipped. From the 1970's onward, big firms began *disassembling* the production chain and redistributing it over several scattered structures, belonging to the same company, or, more generally, belonging to other international firms or local producers system. As a matter of fact, the productive chain changed shape thereafter and it changed in such a fashion that the complexity of the structure increased. It can be shown, by means of graph theory, that the post-fordian structure increased in connectivity and path-cycles diversity across its nodes (Ruzzenenti and Basosi, 2008b). Hence the post-fordian structure presents a higher degree of freedom and thus relates to a more complex system. A system with a higher degree of freedom is a more complex system in the sense that, as for any physical system, it has increased multiplicity or number of different available states. In other words, a more complex system has more ways to arrange the components, in this case goods or raw materials, and therefore to dissipate energy.

According to the hypothesis here advocated, complexity increases when the system can rearrange its components in such a manner that the number and the path length, or the speed of interactions, will be augmented within the same boundaries[18]. That is to say, complexity growth consists of an *intensive* rather than an extensive change, affecting the

---

[18] It noteworthy that, in a network, the number and the path length must be considered intensive features as they can grow without affecting the extension of the network, which is determined eventually only by the number of nodes (components).

internal structure of the system, which may be expressed by a new arrangement of system components[19]. This hypothesis expresses view of complexity based on the concept of geometry[20].

In our opinion the network structure development, that eventually results in complexity growth at any system level, is the outcome of forces (energy influx driven by autocatalytic processes) in the context of hindering boundary conditions. It is the simple growth (in extent and in number of components) the normal behavior. That is to say, without hindering boundary conditions, the system expands its structure, qualitatively unaltered (spatial growth). The system develops in a primary and spatial manner initially, then, when saturation is reached, in a secondary and geometrical (structural) one. It is such geometrical development that enables the system to increase its degree of freedom and to host more energy (or energy density rate) within the same constraints. When this complexity change emerges, the incoming structure, albeit already available to system components, becomes now more probable. The boundary conditions ultimately determine the likelihood of the new structure. It is therefore the role played by *saturation* in system's growth that must be addressed in order to understand the surge of complexity leap.

After the first oil crisis, worldwide industrial production dramatically decreased. There are many clues, indeed, that industrial production at that time reached a saturation point. Statistics show that between the early 1970's and 1990's, a revolution occurred in economic and societal structure that might be considered the end of the industrial age (IEA, 1997). Until that time, linear growth lasted for about 20 years and consisted of a shift in the active population (which was itself growing) from agriculture to industry. It was the nature of the "economic boom", the linear "growth in extent and numbers" (Lotka, 1956). The birth rate thereafter inverted its trend (also in relation to the average income) and population employed in industry reached a maximum and started decreasing (Ruzzenenti and Basosi, 2009b). Industry received a dramatic set back and consequently began to explore new strategies to reduce labour costs and regenerate production. The structural change we have been describing thus far—the globalization and outsourcing revolution—took between 10 and 15 years to become established and influential. However, after the 1990's, the growth trend in the industrial sector resumed and the economy retrieved.

5.2 Degree of freedom reduction/increase

When analyzing the structural complexity change resulting from globalization, of paramount importance is the shift from a uni-located, national productive chain to one that is pluri-located and international. For those firms relying on external resources to pursue their productive needs, production became less costly, but more subdued due to uncontrollable factors. Part of its activity, formerly controlled managerially and internally,

---

[19] An example in cells is represented by the internal skeleton of microtubules that increase the speed of molecules across the cell compared to a transportation system based on simple diffusion. In ecosystems, furthermore, there are food chains and predator-prey dynamics that represent another "transportation network" over which matter flows faster.

[20] According to Lotka, *geometry* is a prominent feature of thermodynamics of living systems and thus, of a sort completely different from those normally addressed by equilibrium thermodynamics. Whereas the latter mainly deals with "structure-less systems", of the like of chemical coefficients, the former must deals with the "geometrical features" of the system (Lotka, 1956).

was then focused on the free market. This shift reduced the stability of the firm and reduced its degree of freedom (choices of allocation). After globalization, firms could explore labour costs according to various national legislations and average incomes. The same occurred for financial and fiscal conditions or the proximity to productive districts. The system (entire market) could thereby reduce production costs by selecting where to set plants or rely on suppliers. It is in this sense that globalization produced the rise of new spatial gradients in the productive system. The whole system thereby increased its degree of freedom. We face therefore the counteractive interplay of degree of freedom, on two hierarchical levels, triggered by a saturation stage. To better stress how this interplay of degree of freedom, working in opposite directions, can be caused by saturation, it is best to approach physical systems for analogies.

For example, if we increase the pressure of a gas in a specific volume, we reduce its degree of freedom and it consequently can become a liquid, at certain temperatures. At the same time, when a liquid changes its motion regime, as in Benárd cells, from a pure, random dissipative system (Figure 5A) to a global dissipative one (Figure 5B), which displays features several magnitudes larger than molecules, a superstructure can arise that was previously available, but very unlikely.

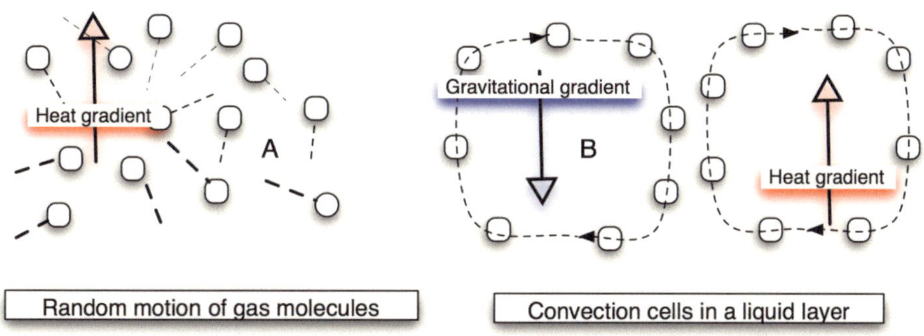

Fig. 5. Degree reduction/increase in dissipative structures (Ruzzenenti and Basosi, 2009b)

Gravity and viscosity constraints make such a structure, beyond a certain level of energy input, possible. The Benárd cells phenomenon is indeed possible when the gradient temperature and water level thickness are known, but not when the vessel permits the fluid to dissipate heat in random motions. In other words, the boundary conditions together with the pressure imposed upon the system by an increasing energy flux, changes the macrostate (energy density) of the system by modifying its microstates (the molecular motion). The random motion of molecules reflects one gradient, the temperature, which is not spatial (geometrical), while the superstructure is exposed to the spatial gradient. That is to say, while the first gradient is defined by one variable, the latter is described by three variables and probabilities consequently change. Dissipation into one variable is therefore more probable than onto three variables, unless boundary conditions render the former impossible. In Benárd cells, such conditions are exemplified in Van der Waals forces, the low heat capacity of water, and restrained vessel thickness (Chandrasekhar, 1961; Prigogine and Stengers, 1984; Swenson, 1997). The connectivity recasts the same trade off in a network system's conceptual framework. A network system grows in complication as long as a new component is connected on the same hierarchical level and it grows in complexity when a new component is introduced on a higher hierarchy (Allen and Starr, 1982). The emergence of a new hierarchy involves coherent behavior for lower level components to the same extent as molecules in Benárd cells, and most importantly, the onset of a new spatial gradient for the higher component, which must now recognize system boundaries. On a molecular scale, cells in the body behave like a network. From the stand point of the organism, however, they act as a whole unit. Indeed "free" cells in substrates are mainly exposed to chemical gradients (temperature, pressure and gravitational gradients as well), while "embedded" cells in tissues that form organs are described by spatial, three dimensional, gradients.

5.3 Spatial symmetry rupture

We believe economic systems (and macroscopic complex systems in general) can exhibit a similar evolutionary pattern: a space symmetry rupture emerges from compelling boundary conditions and increasing energy inflow. In the case of the productive structure's evolution it can be shown that space was isotropic[21] in the former state (fordian) and non isotropic in the latter (post-fordian): a spatial symmetry breaking occurred (Figure 6). What made this spatial gradient rise was a reduction in firms' degree of freedom production settings, coupled with an energy efficiency leap. More energy was thus available to the system amid a condition of hindering forces applied to its boundaries. Two counteractive forces are beneath a symmetry rupture. In this case the symmetry rupture put a space gradient upon the system, with which it induces its variables (components) to organize themselves. *Globalization* and *outsourcing* set production plants in a new, oriented space that was formerly homogeneous.

We would like now to clarify the reason why it has been used the word *rupture* has been used in place of *breaking* to describe the symmetry change. The concept of "symmetry breaking" applies to the temporal scale, whereas here space symmetry has been considered. That is to say, the time-symmetry concept concerns a sudden change in the developmental path of the system; nevertheless this change affects the system itself, rather than the space of the system. In Prigogine's paradigm a dynamic system is considered and it is thus described by a dynamic function, whereas, in the symmetry rupture a *phase transition* rather than a dynamical, however non-linear, change is described[22].

---

[21] In other words, there is just one way to go from the periphery to the centre, regardless of the number of nodes considered, while there are many ways to connect the same number of points in the path. Furthermore, the number of different ways increases with the number of points. This does not mean that, in a scattered productive chain, factories (points), are connected randomly, but instead means that there are multiple ways for a chain to develop its pattern and just one for a centralized system.

[22] „We see therefore, that the appearance of a periodic reaction is a time-symmetry breaking process exactly as ferromagnetism is a space-symmetry breaking one. [….] To understand at least qualitatively this result let us consider the analogy with phase transitions. When we cool down a paramagnetic substance, we come to the so-called Curie point below which the system behaves like a Ferro magnet. Above the Curie point, all directions play the same role. Below, there is a privileged direction corresponding to the direction of magnetization "(Prigogine, 1977).

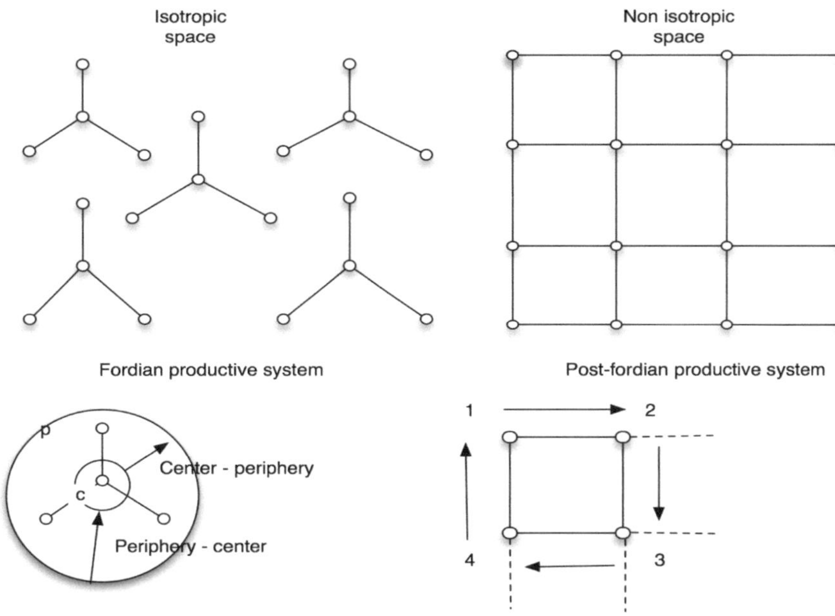

Fig. 6. Spatial symmetry and productive chain (Ruzzenenti and Basosi, 2009b)

It is noteworthy that Prigogine used the concept of space-symmetry breaking as a metaphor to introduce the new concept of time-symmetry breaking. Now, we want to retrieve the concept of space symmetry breaking (symmetry *rupture* in the jargon so far adopted) as we think it is fundamental to understand how evolution may concern the *space* of the system, rather than the system itself. Furthermore, it should also be noticed that the concept of space-symmetry breaking includes the concept of time-symmetry breaking and not *vice versa.*

5.4 Time scale and Spatial scale

Complex systems display a spatial gradient which is sometimes many orders of magnitude larger than gradients involving the scale of components. This important feature of complexity was first envisaged by Prigogine. Parameters describing dissipative structures, like Bénard cells, are macroscopic compared to parameters describing structures at thermodynamic equilibrium. Indeed, while crystals are described by interactions of the order of 10^{-10} meter, convective cells display a size of the order of the 10^{-2} meter (Prigogine and Stengers, 1984). The same can be said for the characteristic time of phenomena. Time scale varies greatly for the above mentioned systems: the vibration period of molecules is of the order of 10^{-15} seconds whereas convective motions have a period in the order of seconds. It is noteworthy that this *scale effect,* consequential to the hierarchical leap, seems to be a common feature of all complex systems, spanning from simple, non-living dissipative structures to greatly complex human-made and biological systems. In the case study here presented the *magnification* of time scale is clearly established in the nature of decision process that characterize the view-point of firms. As long as the entire chain was engulfed in the same firm, if not in the same production plant, any decision inherent to the volume of the production was almost readily attainable. Outsourcing, conversely, brought the productive chain outside the firm, making the *setting* of the plant an endogenous variable and the *volume* of the production an exogenous variable (or at least a less controllable one). Nevertheless, decisions having to do with production's settlement develop themselves over a much larger range of time.

Obviously, the time lag scales up due to the spatial extent of interactions, which increases many orders of magnitude throughout hierarchy as components' size remains the same. This is why the time scale is commensurate to the spatial extent of the system. Nevertheless it seems that magnification of the time scale is affected by space in a fashion that is not entirely reducible to an extensive factor. Time scale as it grows displays a cyclic phenomenon which seems to relate to the symmetry property of the space. In mechanics a cyclic system displays properties of invariance that are proportional to the *symmetry properties* of the space: to every local symmetry there corresponds a conservation law. A conservation law states that some quantity describing a system remains constant throughout its motion; expressed mathematically, the rate of change of its derivative with respect to time is zero.

A system that is cyclic exhibits symmetry as if the space was homogeneous. Therefore by means of cyclic patterns, symmetry in space is re-established and growth can develop again in a continuous way. It is needless to emphasize that cycles are a prominent feature of complex systems, regardless of the nature or the scale that is involved.

6. Conclusions

In the first part of the chapter the rebound effect -the growing energy use coupled with an efficiency enhancement, was employed to introduce the broader question that relates energy efficiency to energy density rate. It was shown that the paradox partially derives from misconception of energy efficiency and power. It must be firstly conceptually set apart *thermodynamic* efficiency from other forms of efficiency. It must than bore in mind that thermodynamic efficiency is strictly connected to power, in two ways. On the one hand there is power-efficiency trade-off and evolutionary systems tend to maximize power rather than efficiency. On the other hand, an efficiency enhancement normally leads to a power shift, as a side effect. In the second part of the chapter, we approached the question of the interdependence between efficiency, complexity and energy density, to illustrate how the causality chain can be reversed: efficiency leads to an higher energy density rate and eventually, to a complexity leap. A complexity leap that is underlined by a change in the space of the system.

As in a phase transition, space symmetry rupture seems to be an important aspect of complexity change. Symmetry rupture, introducing a new gradient in the system space, allows variables (components) to organize themselves. This new arrangement, on the scale of variables, reduces their degree of freedom, on the scale of the whole system, increases it. The change deeply concerns the structure, therefore the *geometry* of the system. Between the new and the old structure a topological change occurs. In our opinion, the topology of the system has to be addressed with graph theory. Yet, the *transition* between the two phases is still an open question and more research is needed. A formal analysis of it should start from recent acquisitions in the field of network theory (Ruzzenenti, Garlaschelli and Basosi, 2010). However, as we tried to illustrate in this chapter, the new arising structure will be more complex and more energy intensive. Higher energy density rate will be an outcome of a transition that will maximize *links* and *frequency* of interactions. Such a transition lays behind, in our opinion, the so called "rebound effect" (Jevons paradox) and explains why energy efficiency has always led to energy growth. Energy conservation policies should therefore contemplate, together with strategies prompted at fostering energy efficiency, measures directed at balancing the long term positive effect over energy consumptions due to a structural changes in economy.

7. References

Alcott B., 2008. The sufficiency strategy: Would rich-world frugality lower environmental impact? Ecological Economics, 6 4, 7 7 0 – 7 8 6

Allen T. H. F., and T. B. Starr. 1982. Hierarchy: perspectives for ecological complexity. University of Chicago Press, Chicago, IL.

Ayres R. U., Warr B., 2005. Accounting for growth: the role of physical work. Structural Change and Economic Dynamics ,16, 181–209

Banks F. E., 2007. The political Economy of World Energy: An Introductory Textbook. Singapore and New York, World Scientific.

Bentzen J., 2004. Estimating the rebound effect in US manufacturing energy consumption, *Energy Economics,* **26**, 123-134.

Binswanger M., 2001. Technological progress and sustainable development: what about the rebound effect? Ecological Economics 36 (2001) 119 - 132

Brookes, Leonard, 1990. The greenhouse effect: the fallacies in the energy efficiency solution. Energy Policy 18 (2), 199–201.

Chaisson E., 2001. Cosmic Evolution-The Rise of Complexity in Nature, Harvard University Press, Cambridge, Massachusetts, London, U.K., 2001.

Chandrasekhar, S. 1961. Hydrodynamic and hydromagnetic stability. Oxford, Clarendon.

Curzon F., Ahlborn B., 1975. Efficiency of a Carnot Engine at Maximum Power Output. American Journal of Physics 41, 22-24 (1975).

Dimitropoulos J., 2007. Energy productivity improvements and the rebound effect: An overview of the state of knowledge. Energy Policy 35, 6354–6363.

Geening, Lorna A., Greene, David L., Difiglio, Carmen, 2000. Energy efficiency and consumption—the rebound effect—a survey. Energy Policy 28 (6/7), 389 - 401.

Grepperud S., Rasmussen I. 2003. General equilibrium assessment of rebound effects, Energy Economics, 2003. 189–203.

Herring H., Energy efficiency—a critical view. Energy 31 (2006) 10–20.

IEA, Indicators of energy use and efficiency, 1997.

IEA Energy Statistics Statistics on the Web: http://www.iea.org/statist/index.htm

Jevons, W. Stanley, The coal question - An Inquiry Concerning the Progress of the Nation, and the Probable Exhaustion of our Coal-mines, M.A., ll.D., F.R.S., Augustus M. Kelley Publisher, New York, 1965

Khazzoom, J.D., 1980, "Economic Implications of Mandated Efficiency in Standards for Houshold Appliances." The Energy Journal, Vol.1, No.4, pp21-40.

Kummel, R., 1989. Energy as a factor of production and entropy as a pollution, Ecological Economics, 1 (1989) 161-180.

Lotka A., 1956. Elements of Mathematical Biology, Dover Publications, Inc, New York. (first publication: elements of physical biology, The Williams and Wilkins Co., Inc, 1924).

Odum E.P., 1997. Ecology: A Bridge Between Science and Society, Sinauer Associates, Inc., Publishers. Sunderland, Massachusetts 01375 U.S.A.

Odum H.T., 1955. The speed regulation: the optimum efficiency for maximum power output in physical and biological systems. American Scientist. 43 (1955), 331-343.

Odum H.T., 1983. Maximum Power and Efficiency: a rebutal, Ecological Modelling, 20, 71-82.

Odum H.T., 1996. Environmental Accounting. Emergy and Environmental Decision Making. John Wiley & Sons, Inc., New York, USA.

Prigogine I., G. Nicolis, A. Babloyantz, 1972, Thermodynamics of evolution. Physics Today 25 (11).

Prigogine Y., 1977. Time, Structure and Fluctuations, Nobel Lecture.

Ruzzenenti, F., Basosi, R., 2008a *The role of the power/efficiency misconception in the rebound effect's size debate: Does efficiency actually lead to a power enhancement?* Energy Policy 36-9, September 2008, 3626-3632.

Ruzzenenti F., Basosi R., 2008b *The rebound effect: An evolutionary perspective.* Ecological Economics, 67 (2008) 526 - 537.

Ruzzenenti F., Basosi R., 2009b *Complexity change and space symmetry rupture,* Ecological Modelling, 220(2009)1880–1885.

Ruzzenenti F., Basosi R., 2009a *Evaluation of the energy efficiency evolution in the European road freight transport sector.* Energy Policy, 37 (2009) 4079–4085.

Ruzzenenti, F., Garlaschelli D., Basosi R., *Complex Networks and Symmetry: a Review with Applications to the Evolution of World Trade. Article in Press, pre-print:* arXiv:1006.3923v1 *[q-fin.GN]*

Saunders, Harry D., 2000. A view from the macro side: rebound, backfire and Khazzoom–Brookes. Energy Policy 28 (6/7), 439–449.

Schroeder D., 2000. Thermal Physics, an introduction to. Addison Wesley Longman.

Schipper L., Haas R., 1998. Residential energy demand in OECD-countries and the role of irreversible efficiency improvements. Energy Economics 20 (1998) 421-442.

Schipper L., Grubb M., 2000 On the rebound? Feedback between energy intensities and energy uses in IEA countries. Energy Policy 28, 367 } 388

Sorrell, S., Dimitropoulos, J., 2007. UKERC Review of Evidence for the Rebound Effect: Technical Report 5—Energy Productivity and Economic Growth Studies. UK Energy Research Centre, London.

Sorrell S., 2009. Jevons' Paradox revisited: The evidence for backfire from improved energy efficiency. Energy Policy, 37 (2009) 1456-1469.

Swenson R., 1997. Autokatakinetics, evolution, and the law of maximum entropy production: a principled foundation toward the study of human ecology, in: Freese, L., Advances in human Ecology, vol.&. JAI Press, Greewich, CT, pp. 1-47; 1989. Emergent attractors and the law of maximum entropy production: foundations to a theory of general evolution. Syst. Res. 6, 187-197.

Washida T., 2004. Economy-wide Model of Rebound Effect for Environmental Efficiency , proceedings of International Workshop on Sustainable Consumption, University of Leeds, March 5-6, 2004.

3

Categorizing Barriers to Energy Efficiency: An Interdisciplinary Perspective

Patrik Thollander[1], Jenny Palm[2] and Patrik Rohdin[1]
[1]*Energy systems, Linköping University*
[2]*Tema T, Linköping University*
Sweden

1. Introduction

This chapter presents theoretical perspectives on barriers to energy efficiency identified in different scientific disciplines, and briefly describes each barrier and its mode of operation. In an attempt to categorize barriers to energy efficiency, the chapter addresses socio-technical regimes, leading to a novel interdisciplinary categorization of barriers to energy efficiency in *three categories*: the technological system, the technological regime, and the socio-technical regime.
The threat of climate change resulting from the use of fossil fuels is posing a threat to the environment, and energy efficiency is one of the most important means of reducing this threat (IPCC, 2007). Despite this, there are a number of publications stating the existence of a "gap" between potential cost-effective energy efficiency measures and measures actually implemented—the so called "energy efficiency gap" or "energy paradox" (York et al., 1978; Blumstein et al., 1980; Stern and Aronsson, 1984; Hirst and Brown, 1990; Gruber and Brand, 1991; Stern, 1992; DeCanio, 1993; Jaffe and Stavins 1994; Sanstad and Howarth, 1994; Weber, 1997; Ostertag, 1999; Sorrell et al., 2000; Brown, 2001; de Groot et al., 2001; Schleich, 2004; Sorrell et al., 2004; Schleich and Gruber, 2008). This "energy efficiency gap" or "energy paradox" exists due to barriers to energy efficiency. A barrier may be defined as a postulated mechanism that inhibits investments in technologies that are both energy-efficient and economically efficient (Sorrell et al., 2004).
Barriers are explanations for the reluctance to adopt cost-effective energy efficiency measures derived from mainstream economics, organizational economics, and organizational and behavioural theories. There are also institutional or structural barriers to energy efficiency that do not directly affect the "gap", even though it does affect the overall level of energy efficiency. Barriers may be divided into three broad categories: *Economic*, *Organizational* and *Behavioural*. Inspired by an extensive review of the existing literature on barriers to energy efficiency, Sorrell et al. (2000) compiled a barrier framework categorized into different barriers (see Table 1; the barriers are explained in greater detail in the following sections of this chapter). It should be noted that the above classification of barriers is not unambiguous; one type of real-world phenomena may be explained by several of the theoretically derived barriers presented (Weber, 1997).

Jaffe and Stavins (1994) outlined a number of different levels of "energy efficiency potential", or "energy efficiency gaps" (see Figure 1). The figure states that the actual potential level of energy efficiency depends on which view is applied—while the technologist's potential is real in a sense, the economist's potential is actually real for that person or organization, with the difference between the two levels depending on which theoretical perspective is being applied.

Theoretical Barriers	Comment
Imperfect information (Howarth and Andersson, 1993)	Lack of information may lead to cost-effective energy efficiency measures opportunities being missed.
Adverse selection (Sanstad and Howarth, 1994) (Jaffe and Stavins, 1994)	If suppliers know more about the energy performance of goods than purchasers, the purchasers may select goods on the basis of visible aspects such as price.
Principal-agent relationships (Jaffe and Stavins, 1994)	Strict monitoring and control by the principal, since he or she cannot see what the agent is doing, may result in energy efficiency measures being ignored.
Split incentives (Jaffe and Stavins, 1994) (Hirst and Brown, 1990)	If a person or department cannot gain benefits from energy efficiency investment it is likely that implementation will be of less interest.
Hidden costs (Jaffe and Stavins, 1994) (Ostertag, 1999)	Examples of hidden costs are overhead costs, cost of collecting and analyzing information, production disruptions, inconvenience etc..
Access to capital (Hirst and Brown, 1990) (Jaffe and Stavins, 1994)	Limited access to capital may prevent energy efficiency measures from being implemented.
Risk (Hirst and Brown, 1990)	Risk aversion may be the reason why energy efficiency measures are constrained by short pay-back criteria.
Heterogeneity (Jaffe and Stavins, 1994)	A technology or measure may be cost-effective in general, but not in all cases.
Form of information (Stern and Aronsson, 1984)	Research has shown that the form of information is critical. Information should be specific, vivid, simple, and personal to increase its chances of being accepted.
Credibility and trust (Stern and Aronsson, 1984)	The information source should be credible and trustworthy in order to successfully deliver information regarding energy efficiency measures. If these factors are lacking this will result in inefficient choices.
Values (Stern, 1992)	Efficiency improvements are most likely to be successful if there are individuals with real ambition, preferably represented by a key individual within top management.
Inertia (Stern and Aronsson, 1984)	Individuals who are opponents to change within an organization may result in overlooking energy efficiency measures that are cost-effective.
Bounded rationality (Sanstad and Howarth, 1994)	Instead of being based on perfect information, decisions are made by rule of thumb.
Power (Sorrell et al., 2000)	Low status of energy management may lead to lower priority of energy issues within organizations.
Culture (Sorrell et al., 2000)	Organizations may encourage energy efficiency investments by developing a culture characterized by environmental values.

Table 1. Classification of barriers to energy efficiency (inspired by Sorrell et al., 2000).

1.1 Economic barriers – market failures

One important category with regard to barriers is the group of barriers that may be seen as market failures violating the underlying axioms of mainstream economic theory. According to mainstream economic theory, a market failure may justify public policy intervention. However, the mere existence of a market failure may not in and of itself be sufficient to justify intervention. As Brown (2001) writes:

"The existence of market failures and barriers that inhibit socially optimal levels of investment in energy efficiency is the primary reason for considering public policy interventions. In many instances, feasible, low cost policies can be implemented that either eliminate or compensate for market imperfections and barriers, enabling markets to operate more efficiently to the benefit of society. In other instances, policies may not be feasible; they may not fully eliminate the targeted barrier or imperfection; or they may do so at costs that exceed the benefits." (Brown, 2001).

The elimination of a market failure barrier may thus only be put into operation if the benefits arising from an intervention exceed the cost of implementation.

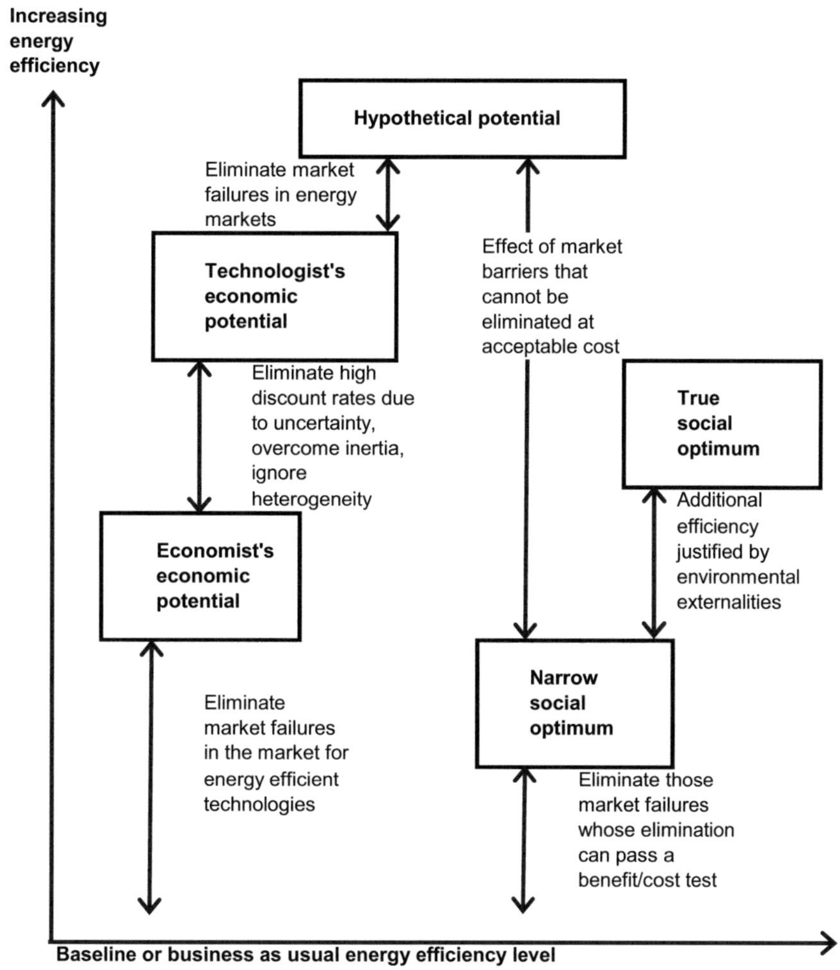

Fig. 1. Different levels of energy efficiency potential (Jaffe and Stavins, 1994).

An often cited market failure barrier is *imperfect information*. Other market failure barriers include *asymmetric information*, a special form of imperfect information where *split incentives*, *adverse selection,* and *principal-agent relationships* may also be categorized. These market failure barriers are presented below.

1.1.1 Imperfect information

A large body of research states that consumers are often poorly informed about market conditions, technology characteristics and their own energy use. The lack of adequate information about potential energy-efficient technologies inhibits investments in energy efficiency measures (Sanstad and Howarth, 1994). Insufficient information is one form of imperfect information, such as when the energy performance of energy-efficient technologies is not made available to agents. Another form of imperfect information is the cost of information, meaning that there are costs associated with searching and acquiring information about the energy performance of an energy-efficient technology. Yet another form is the accuracy of information, meaning that the information provider may not always be transparent about the product being offered. Imperfect information is likely to be most serious when the product is purchased infrequently, performance characteristics are difficult to evaluate either before or soon after purchase, and the rate of technology change is rapid relative to the purchase intervals (Sorrell et al., 2000), which is the case for many energy efficiency measures. Issues related to imperfect information may be countered with different forms of information campaigns.

1.1.2 Adverse selection

Adverse selection means that producers of energy-efficient equipment are, in general, more informed about the characteristics and performance of equipment than prospective buyers. In other words, the information between the two parties engaged in the transaction is asymmetric. Since asymmetric information is extremely common in real world markets, inefficient outcomes may be the rule rather than the exception (Sanstad and Howarth, 1994).

1.1.3 Principal-agent relationship

The principal-agent relationship arises due to a lack of trust between two parties at different levels within an organization or transaction. The owner of a company, who may not be as well-informed about the site-specific criteria for energy efficiency investments, may demand short payback rates/high hurdle rates on energy efficiency investments due to his or her distrust in the executive's ability to convey such investments—leading to the neglect of cost-effective energy efficiency investments (DeCanio 1993; Jaffe and Stavins, 1994).

1.1.4 Split incentives

A split incentive may occur when the potential adopter of an investment is not the party that pays the energy bill. If so, information about available cost-effective energy efficiency measures in the hands of the potential adopter may not be sufficient; adoption will only occur if the adopter can recover the investment from the party that enjoys the energy savings (Jaffe and Stavins, 1994). This is often referred to as the landlord-tenant relationship For example, if a mid-level executive pays the energy bill for his or her division based on number of employees, this decreases interest in the organization's overall in-house energy program to lower energy costs (including investments in energy efficiency technologies), since there is "nothing in it" for him or her. This is a restriction to adopting energy-efficient technologies, in particular those with higher initial costs but lower life cycle costs than conventional technologies (Hirst and Brown, 1990). The lack of sub-metering within multidivisional organizations may also be classified as a split incentive.

1.2 Economic barriers: non-market failures

Apart from market failure barriers, there are a number of barriers that explain the "gap" but which cannot be categorized as market failures, but are rather *non-market failure* barriers or market barriers. A market barrier, according to Jaffe and Stavins (1994), may be defined as *any factor that may account for the "gap"*, while Brown (2001) defines market barriers as *obstacles that are not based on market failures but which nonetheless contribute to the slow diffusion and adoption of energy-efficient measures.* Barriers that may be categorized as market barriers are, for example, hidden costs, limited access to capital, risk, and heterogeneity. These barriers are presented below.

1.2.1 Hidden costs

Hidden costs are often used as an explanatory variable for the "gap" (DeCanio, 1998). In short, the argument is that there are high costs associated with information-seeking, meeting with sellers, writing contracts and other such activities; if these costs are higher than the actual profit from implementation, they inhibit investment. Accordingly, cost-effective measures are not cost-effective when such costs associated with the investment are included. A study by Hein and Blok (1994) found that hidden costs in large energy-intensive industrial firms ranged from three to eight percent of total investment costs. In smaller, non-energy-intensive firms, such costs are thus likely to be even higher. Hidden costs are a frequently used argument against the existence of an energy efficiency gap; it is argued that engineering-economic models are not able to see the full cost of an energy efficiency measure (Sorrell et al., 2000).

1.2.2 Limited access to capital

Technologies that are energy-efficient are often more expensive to purchase than alternative technologies (Almeida, 1998). Moreover, obtaining additional capital in order to invest in energy-efficient technology may be problematic. Apart from low liquidity, limited access to capital may also arise due to restrictions on lending money (Hirst and Brown, 1990). Sometimes such restrictions may be self-imposed.

1.2.3 Risk

Even though, for example, managers know what the capital cost is for an energy efficiency investment, there can be uncertainty about the long-term savings in operating costs; this means the investment poses a risk. Such concerns have been found to be very important to decision-makers (Hirst and Brown, 1990).

Stern and Aronson (1984) also identify risk as a barrier to energy efficiency, since accurate estimates of the net costs of implementing energy efficiency measures depend on future economic conditions in general, and on future energy prices and availability in particular. Energy prices have fluctuated as long as there has been a market for energy, leading to perceptions of uncertainty about future prices. *How are consumers to make "rational" choices about the purchase of new energy-using systems such as cars, heating equipment, new buildings, and motors when the basis for estimating long-term operating costs is so uncertain? ... Uncertainty about fuel prices is a barrier to investment in both the manufacture and purchase of energy-efficient systems* (Hirst and Brown, 1990). Studies among small and medium-sized enterprises have found that some may not even be able to reduce uncertainty to a calculated risk due to a lack of time and money to calculate the required estimates (Stern and Aronson, 1984).

1.2.4 Heterogeneity

The heterogeneity barrier is associated with the fact that even if a given technology is cost-effective on average, it will most likely not be so for some individuals or firms. Heterogeneity particularly impacts production processes of companies that often specialize in one type of goods, and where a potential energy efficiency measure may be difficult to implement in another company. Even though similar goods are produced, small differences in the products, such as different size and shape, can inhibit the implementation of the measure in another firm (Jaffe and Stavins, 1994). Heterogeneity may be an explanatory variable for the "gap" when constructing (economic) models of a population of companies, but is less likely to hold if site-specific information exists regarding a cost-effective energy efficiency measure resulting from, for example, an energy audit.

1.3 Behavioural barriers

Apart from the explanations for the "gap" outlined above, there are also a number of barriers derived from behavioural sciences that explain the "gap", such as the form of information, credibility and trust, values, inertia, and bounded rationality. These barriers are presented below.

1.3.1 Form of information

One barrier to energy efficiency is the form of information, meaning that information does not always receive as much attention as anticipated, since people are (often) not active information-seekers but rather selective about attending to and assimilating information. Research points out some characteristics in the way information is assimilated; some people, for example, are more likely to remember information if it is specific and presented in a vivid and personalized manner, and comes from a person who is similar to the receiver (Stern and Aronson, 1984; Palm, 2009, 2010).

1.3.2 Credibility and trust

Another factor that may inhibit adoption is the receiver's perceived credibility of and trust in the information provider. Energy users cannot always easily gain accurate information about the ultimate comparative cost of different investment options; they will rely on the most credible available information. The following example from the household sector may illustrate this. Pamphlets describing how to save energy in home air conditioning systems were sent out to 1,000 households in New York. Fifty percent of the households received the information in a mailing from the local electricity utility, and the other half received it from the state regulatory agency for utilities. The following month, households that had received the pamphlet from the state agency used about eight percent less electricity than the households that had received the same pamphlet from the local electricity utility (Stern and Aronson, 1984). The effective spread of information thus depends on a trustworthy information provider. As regards the industry, intermediaries such as sector organizations or consultants may play an important role, as these entities or individuals often tend to be regarded as trustworthy (Ramirez et al., 2005; Stern and Aronson, 1984).

1.3.3 Values

Values such as helping others, concern for the environment and a moral commitment to use energy more efficiently are influencing individuals and groups of individuals to adopt energy efficiency measures. However, studies of households indicate that norms only have a strong impact on cost-free energy efficiency and energy conservation measures (Stern and Aronson, 1984). A study by Aronson and O'Leary (1983) on showering in a university building showed that the number of students taking short, energy-saving showers increased from six percent when a sign encouraging short showers was put up, to 19 percent when an intrusive sign was used, to 49 percent when the researchers used a student to set an example for others by always turning off the water and soaping up whenever someone came into the facility, and to 67 percent when two students serving as examples were used (Aronsson and O'Leary, 1983). Consequently, a lack of values related to energy efficiency may inhibit measures from being undertaken.

1.3.4 Inertia

In short, inertia means that individuals and organizations are, in part, creatures of habit and established routines, which may make it difficult to create changes to such behaviours and habits. This is stated as an explanatory variable to the "gap". People work to reduce uncertainty and change in their environments, and avoid or ignore problems (Stern and Aronson, 1984). Also, people who have recently made an important decision often seek to justify that decision afterwards—convincing themselves and others that the decision was correct. This description of inertia may partially explain the failure of many energy users to take economically justifiable actions to save energy; energy efficiency also often begins with small commitments that later lead to greater ones (Stern and Aronson, 1984).

1.3.5 Bounded rationality

Another explanation for why cost-effective energy efficiency measures are not undertaken is bounded rationality (Simon, 1957). Most types of market failures are concerned with problems in the economic environment that impede economic efficiency even when assuming fully rational agents—that is, utility-maximizing consumers and profit-maximizing firms (Palm and Thollander, 2010). In the case of energy efficiency-related decisions, this hypothesis formally requires decision-makers to solve what may be extremely complex optimization problems in order to obtain the lowest-cost provision of energy services (Sanstad and Howarth, 1994). Studies of organizational decision-making identify two major features of organizations that affect the linkage of a simple rational view to their actions. First, the organization is not a single actor but rather consists of many actors with different, sometimes conflicting, objectives. The interests of one employee or department may, for example, be in conflict with those of others. Second, according to

Sanstad and Howarth (1994), organizations (just like individuals) to some extent do not act on the basis of complete information but rather make decisions by rule of thumb (Stern and Aronson, 1984).

1.4 Organizational barriers

Apart from economic and behavioural barriers, there are also barriers such as power and culture that emerge from organizational theory. These barriers are presented below.

1.4.1 Power

Lack of power among energy efficiency decision-makers (e.g., the energy controllers), is often put forth as an explanatory variable for the "gap". The low importance of energy management within organizations leads to constraints when striving to implement energy efficiency measures (Sorrell et al., 2000).

1.4.2 Culture

Culture is closely connected to the values of the individuals forming the culture. An organization's culture may be seen as the sum of each individual's values, where the executives' values or the values of other workers who have influence within the organization may have more impact on the organization's culture than "lower status" workers (Sorrell et al., 2000).

1.5 Different ways of categorizing barriers to energy efficiency

A review of research on barriers to energy efficiency reveals that a number of different means of categorizing barriers exists.

A barrier model specifies three features: the objective obstacle, the subject hindered, and the action hindered. The methodological question of how to determine a barrier model is: what is an obstacle to whom reaching what in energy conservation (Weber, 1997)?

- What is an obstacle (persons, patterns of behaviour, attitudes, preferences, social norms, habits, needs, organizations, cultural patterns, technical standards, regulations, economic interests, financial incentives, etc.)
- ... is an obstacle to whom (consumers, tenants, workers, clerks, managers, voters, politicians, local administration, parties, trade unions, households, firms, non-governmental organizations)
- ... reaching what (buying more efficient equipment, retro-fitting, decreasing an energy tax, establishing a public traffic network, improving operating practices, etc.)

Different ways of categorizing barriers to energy efficiency have been developed. Sorrell et al. (2000) distinguish three main categories: market failures, organizational failures and non-failures, while Weber (1997) classifies the barriers as institutional, economic, organizational and behavioral barriers. Hirst and Brown (1990) made yet another distinction of barriers to energy efficiency, which divides the barriers into two broad categories: structural barriers and behavioral barriers.

In the following section we will discuss another way of understanding technological development and changes in organizations, namely transition theory and socio-technical regimes.

2. Socio-technical regimes

At this stage it is useful to introduce Geels et.al.'s evolutionary model for socio-technical change, which focuses on the dynamics in changing artifacts, technologies, regimes and overall society. The model relies on the work of science and technology studies (STS), which argues that technological and social change are interrelated.

In this model, radical novelties are developed in special spaces or technological niches, where they are sheltered from mainstream competition (Schot and Geels, 2008). These can be small market niches or technological niches where resources are provided by public subsidies. Niches need protection because new technologies initially have low price/performance ratios. Since small networks of actors protect the niches, when initiating new technology building social networks is a vital activity (Verbong and Geels, 2007).

Niches form the micro level at which radical novelties emerge. The meso level is the regime level, and includes routines, knowledge, defining problems and so on embedded in institutions and infrastructures (Shove 2003). The macro level is the socio-technical landscape, which is the environment that changes slowly. Verbong and Geels (2007) describe the relationship between the three levels as a "nested hierarchy". New technologies have problems breaking through because of deep-rooted, established regimes. Transition only takes place when all three levels link up and reinforce each other.

Geels (2004) has developed Nelson and winter's "technological regimes" and discusses socio-technical regimes. Technological regimes refer to cognitive routines that are shared in a community of engineers and that guides research and development activities. The technological regime is the rule-set embedded in "engineering practices, production process technologies, product characteristics, skills and procedures, ways of handling relevant artefacts and persons, ways of defining problems; all of them embedded in institutions and infrastructures". It highlights the fact that engineers act in a social context of social structures, regulations and norms (Geels and Kemp, 2007, pp 443). Technological regimes are broadened to include socio-technical regimes by including the institutional and market aspects needed to make the technical regime work. A socio-technical regime is characterized by the set of rules that guide technical design, as well as the rules that shape market development such as user preferences and rules for regulating these markets (Schot and Geels, 2007). The use of socio-technical regimes also implies the existence of different regimes and the existence of a connection and mutual dependency between them. In a company, different social groups can be distinguished by their own special features. Actors within these groups then share a set of rules, or a regime. Because different groups share different rules, it is possible to distinguish different regimes, such as technological regimes, science regimes, and financial regimes and so on. They share aims, values, problems, agendas, professional journals, etc. However, rules are not just linked within regimes but also between regimes, and regimes influence each other; this is why socio-technical regimes are a better concept for explaining this (Geels, 2004). When regimes are widened to socio-technical regimes, they include interaction with other social groups, besides engineers and firms, in society such as users, policy-makers and social groups. Regimes not only refer to cognitive routines and belief systems, but also to regulative rules and normative roles. From this perspective, different regimes are relatively autonomous, but also interdependent. A socio-technical regime thus binds producers, users and regulators together.

As mentioned above, the socio-technical regime forms the meso level, which accounts for the stability of existing large-scale systems such as energy systems. The macro level is formed by the socio-technical landscape, and cannot be under direct influence of niche and regime actors. Changes at the landscape level occur slowly. Niche actors hope that novelties will eventually be used in the regime. Niche actors can contribute to changes in the practices and routines of existing regime actors. Sometimes niches can also replace the existing regime. It is not easy, however, to replace an established regime, not least because of lock-in effects wherein new technology often needs to fit into existing system solutions (Schot and Geels, 2008).

Socio-technical regimes highlight the fact that actors are embedded in structures that shape their preferences, aims and strategies. But from this perspective, actors also have agency and perform conscious and strategic actions. The model confirms Gidden's duality of structure, and when that structure produces and mediates action. Actors can then act upon and restructure these systems (Geels, 2004). Regimes then implement and (re)produce rules in social activities that take place in local practices. By implementing shared rule systems, the regime actors generate patterns of activity that are similar across different local practices. There may be variation, however, between local practices due to the fact that there are differences between group members, so regimes can have somewhat different strategies, resources, problems and aims. Strategies, aims and the like are also not very flexible within a regime, and undergo only incremental change over time (Geels, 2004). In addition, incremental innovation still occurs in stable regimes and is important because these changes can accumulate and result in major performance improvements over time (Geels and Kemp, 2007).

A dominant regime can be forced to restructure and invest in new technical directions. For example, changes in the socio-technical landscape can put pressure on the regime. Climate change has forced the energy and transport sector to find new technical strategies. Internal technical problems, change in user preferences and negative externalities such as health risks may also trigger actors to act. Competitive games between firms are another example of developments that can open up a regime (Geels, 2004).

If we cross-pollinate barriers theories with ideas from transition theories and socio-technical regimes, we have a new categorization of barriers and, therefore, a new way of reflecting on and discussing efficiency gaps. This will be discussed in the following section.

3. Conclusions: A proposed structure for empirical studies on barriers to energy efficiency

How we define a problem determines whether we can solve it; this is elementary knowledge in all of the sciences. Clear definitions are the foundation for all innovative thoughts, which is why it is important to discuss how barriers to energy efficiency can be categorized in potentially different ways. In an attempt to categorize barriers to energy efficiency, the 15 theoretical barriers are divided into three different categories, depending on each barrier's system complexity (see table 2). In the first category—the technical system—the results are quite restricted to technology and its associated costs. In the second category—the technological regime—the results are influenced by human factors but nevertheless coupled to the technology in question. In the third category—the socio-technical regime—the results are heavily influenced by human factors, and less influenced by the technology in question.

Classification	Theoretical Barriers
The technical system	Access to capital (Hirst and Brown, 1990)
	Heterogeneity (Jaffe and Stavins, 1994)
	Hidden costs (Ostertag, 1999)
	Risk (Hirst and Brown, 1990)
The technological regime	Imperfect information (Howarth and Andersson, 1993)
	Adverse selection (Sanstad and Howarth, 1994)
	Split incentives (Jaffe and Stavins, 1994)
	Form of information (Stern and Aronsson, 1984)
The socio-technical regime	Credibility and trust (Stern and Aronsson, 1984)
	Principal-agent relationship (Jaffe and Stavins, 1994)
	Values (Stern, 1992)
	Inertia (Stern and Aronsson, 1984)
	Bounded rationality (Sanstad and Howarth, 1994)
	Power (Sorrell et al., 2000)
	Culture (Sorrell et al., 2000)

Table 2. Proposed classification of barriers to energy efficiency.

Re-defining how we should categorize barriers could open up new ways of looking at the problem, which in turn might lead to other suggestions for addressing the energy efficiency gap. Energy efficiency problems are multi-faceted and should be approached accordingly. If a barrier is identified as belonging to a technological regime or a socio-technical regime, it should be approached differently and addressed via different policy means. If a barrier is seen as belonging to a technological regime, then more information on existing energy efficient measures could be a possible solution. If a barrier is more related to a socio-technical perspective on barriers, then aspects such as corporate culture and established

internal values should be problematized and highlighted. In other words, how we perceive and define these barriers will lead to different solutions for overcoming the barriers and, ultimately, to different policy recommendations.
Finding solutions to the energy efficiency gap is vital for solving the climate change problem. To define and redefine the empirically identified barriers is therefore important for challenging existing solutions and developing new, creative ways of approaching companies and other actors. Employing this categorization of barriers would lead to a greater focus on social practices in companies and existing routines in decision-making and industrial processes.

4. References

Almeida, E. L. (1998). Energy efficiency and the limits of market forces: The example of the electric motor market in France. *Energy Policy*, 26, 8, 643–653, ISSN 0301-4215.

Aronson, E., O'Leary, M. (1983). The relative effectiveness of models and prompts on energy conservation: field experiment in a shower room. *Journal of Environmental Systems*, 12, 3, 219-224, ISSN 0047-2433.

Blumstein, C., Krieg, B., Schipper, L., York, C.M. (1980). Overcoming social and institutional barriers to energy conservation. *Energy*, 5, 355-371, ISSN 0144-2600.

Brown, M.A. (2001). Market failures and barriers as a basis for clean energy policies., *Energy Policy*, 29, 14, 1197-1207, ISSN 0301-4215.

de Groot, H., Verhoef, E., Nijkamp, P. (2001). Energy saving by firms: decision-making, barriers and policies. *Energy Economics*, 23, 6, 717-740, ISSN 0140-9833.

DeCanio, S. (1998). The efficiency paradox: bureaucratic and organizational barriers to profitable energy-saving investments. *Energy Policy*, 26, 5, 441-458, ISSN 0301-4215.

DeCanio, S. (1993). Barriers within firms to energy efficient investments. *Energy Policy*, 9, 1, 906-914, ISSN 0301-4215.

Geels, F. (2004) From Sectoral systems of innovation to socio-technical systems. Insights about dynamics and change from sociology and institutional theory. *Research policy*, 33, 897-920, ISSN 0048-7333.

Geels, F and Kemp, R. (2007). Dynamics in socio-technical systems: Typology of change processes and contrasting case studies. *Technology in Society*, 29, 441-455, ISSN 0160-791x.

Gruber, E., Brand, M. (1991). Promoting energy conservation in small and medium-sized companies. *Energy Policy*, 19, 3, 279-287, ISSN 0301-4215.

Hein, L., Blok, K. (1995). Transaction costs of energy efficiency improvement. In Proceedings of the 1995 ECEEE summer study, Panel 2, 1-8.

Hirst, E., Brown, M., A.(1990). Closing the efficiency gap: barriers to the efficient use of energy. *Resources, Conservation and Recycling*, 3, 4, 267-281, ISSN 0921-3449.

Howarth, R., Andersson, B. (1993). Market barriers to energy efficiency. *Energy Economics*, 15, 4), 262-272, ISSN 0140-9833.

Jaffe, A.B., Stavins, R.N. (1994). The energy efficiency gap: what does it mean? *Energy Policy*, 22, 10, 60-71, ISSN 0301-4215.

Ostertag, K. (1999). Transaction Costs of Raising Energy Efficiency. In: Proceedings of the 2007 IEA international Workshop on Technologies to Reduce Greenhouse gas Emissions: Engineering-Economic Analyses of Conserved Energy and Carbon. Washington DC, 5-7 May 1999.

Palm, J. (2009). Placing barriers to industrial energy efficiency in a social context: a discussion of lifestyle categorisation. *Energy Efficiency*, 2, 3, 263-270, ISSN 1570-646x.

Palm, J. (2010). The public-private divide in household bahavior. How far into the home can energy guidance reach? *Energy Policy*, 38, 6, 2858-2864, ISSN 0301-4215.

Palm, J. and Thollander, P. (2010). An interdisciplinary perspective on industrial energy efficiency. *Applied Energy* 87, 10, 3255-3261, ISSN 0306-2619.

Ramirez, C.A., Patel, M., Blok, K. (2005). The non-energy intensive manufacturing sector. An energy analysis relating to the Netherlands. *Energy*, 30, 5, 749-767, ISSN 0144-2600.

Rohdin, P., Thollander, P. (2006). Barriers to and driving forces for energy efficiency in the non-energy-intensive manufacturing industry in Sweden, *Energy* 31, 12, 1836-1844, ISSN 0144-2600.

Rohdin, P., Thollander, P., Solding, P., 2007. Barriers to and drivers for energy efficiency in the Swedish foundry industry. *Energy Policy* doi: 10.1016 35, 1, 672-677, ISSN 0301-4215.

Sanstad, A., Howarth, R.,(1994). 'Normal' markets, market imperfections and energy efficiency. *Energy Policy*, 10, 811-818, ISSN 0301-4215.

Schleich, J., Gruber, E. (2008). Beyond case studies: Barriers to energy efficiency in commerce and the services sector. *Energy Economics*, 30, 2, 449-464, ISSN 0140-9833.

Schleich, J. (2004). Do energy audits help reduce barriers to energy efficiency? An empirical analysis for Germany. *International Journal of Energy Technology and Policy*, 2, 3, 226-239, ISSN 1472-8923.

Schot, J and Geels, F. (2007). Niches in evolutionary theories of technical change. A critical survey of the literature. *Journal of Evolutionary Economics*, 17, 605-622, ISSN 0936-9937.

Schot, J and Geels, F. (2008) Strategic niche management and sustainable innovation journeys: theory, findings, research agenda and policy. *Technology Analysis & Strategig Management*, 20, 5, 537-554, ISSN 0953-7325.

Shove, E. (2003). Users, Technologies and Expectations of Comfort, Cleanliness and Convenience. *Innovation*, 16, 2, 193-205, ISSN 1469-8412.

Simon, H.A. (1957). *Models of Man*. Wiley, London.

Sorrell S., O'Malley, E., Schleich, J., Scott, S. (2004). *The Economics of Energy Efficiency - Barriers to Cost-Effective Investment*, Edward Elgar, Cheltenham.

Sorrell, S., Schleich, J., Scott, S., O'Malley, E., Trace, F., Boede, E., Ostertag, K. Radgen, P. (2000). Reducing Barriers to Energy Efficiency in Public and Private Organizations. Retrieved October 8, 2007, from the SPRU's (Science and Technology Policy Research) Retrieved October 8, 2007, from: http://www.sussex. ac.uk/Units/spru /publications/reports/ barriers/final.html.

Stern, P.C. (1992). What Psychology Knows About Energy Conservation. *American Psychologist*, 47, 10, 1224-1232, ISSN 0003-066x.

Stern, P.C., Aronson, E. (1984, Eds). *Energy Use: The Human Dimension*, W.H Freeman, 0716716216, New York.

Thollander, P., Ottosson, M., 2008. An energy-efficient Swedish pulp and paper industry – exploring barriers to and driving forces for cost-effective energy efficiency investments. *Energy Efficiency* 1, 1, 21-34, ISSN 1570-646x.

Thollander, P., Rohdin, P., Danestig, M., 2007. Energy policies for increased industrial energy efficiency: Evaluation of a local energy programme for manufacturing SMEs. *Energy Policy* 35, 11, 5774-5783, ISSN 0301-4215.

Verbong, G and Geels, F. (2007). The ongoing energy transition: Lessons from a socio-technical multi-level analysis of the Dutch electricity system (1960-2004). *Energy Policy*, 35, 1025-1037, ISSN 0301-4215.

Weber, L. (1997). Some reflections on barriers to the efficient use of energy. *Energy Policy*, 25, 10, 833-835, ISSN 0301-4215.

York, C.M., Blumstein, C., Krieg, B., Schipper, L. (1978). *Bibliography in institutional barriers to energy conservation*. Lawrence Berkeley Laboratory and University of California, Berkeley.

4

Factors influencing energy efficiency in the German and Colombian manufacturing industries

Clara Inés Pardo Martínez
University of Wuppertal, Wuppertal Institute and University of La Salle
Germany and Colombia

1. Introduction

Energy is a basic factor for industrial production, and the level of electricity consumption is used to measure the progress and economic development of nations. Globally, growing population, industrialisation and rising living standards have substantially increased dependence on energy. As a result, the development of conventional energy resources, the search for new or renewable energy sources, energy conservation (using less energy), and energy efficiency (same service or output, less energy) have become unavoidable topics within politics.

Generally, an ideal policy cycle sees a given policy formulated, implemented, monitored and evaluated to verify its effectiveness and fulfilment of the proposed objectives and in accordance with the results of this evaluation, the policy is then kept, reformulated or abolished. In this cycle—and above all, in industrial energy politics—it is important that the policy makers recognise the influence of economic, technical and political factors and have an understanding of the mechanisms that determine energy efficiency performance such that the instruments and strategy they formulate become successful.

Strategies and instruments developers drafting an energy policy need to understand the behaviour of the manufacturing industry with respect to energy consumption in order to (i) motivate, (ii) target energy actions that will be adopted, and (iii) develop energy saving and energy efficiency actions and technologies that will be of interest (Kant, 1995 and Thollander et al., 2007). The quantity and quality of energy conservation support or energy efficiency programs will depend on perceived interest and as well as the need for energy conservation changes.

There are limited studies and information currently available on the perception of approach to energy efficiency in companies. Therefore, this study seeks to analyse the factors and strategies that address energy efficiency in the manufacturing industries. This information may be useful for energy policy and program development as well as pollution prevention and energy efficiency strategies. The research questions that guide this chapter are:

- What is the role of energy consumption and energy efficiency in business strategies in the manufacturing industries?
- What are the variables of political factors that may have more influence on energy efficiency performance?

- What are the strategies and instruments that may generate better results to improve energy efficiency in the manufacturing industries?

These questions were investigated in this study by means of the opinions and expectations of the main stakeholders (associations and representative firms in Germany and Colombia) through a questionnaire and analysis of literature.

This chapter is structured as follows. In section 2, examines energy efficiency policy in both countries. Section 3 shows the methodology used in this study. Section 4 analyses changes in energy efficiency in German and Colombian manufacturing industries. Results and discussion appear in section 5 while the section 6 shows different strategies and recommendations for an effective energy efficiency policy in the Colombian manufacturing industry. The main conclusions of the study are presented in section 7.

2. General characteristics of energy efficiency policy in Germany and Colombia

2.1 The German energy efficiency policy

The German energy policy is based in the commitment to the "3 Es": energy security, economic efficiency and environmental sustainability. In this context, Germany emphasises environment and climate change objectives, and energy efficiency assumes increased importance in the country's overall energy policy. Moreover, in the last decade, the key German energy policies have been based on the expansion of the use of renewable energy and the establishment of new energy efficiency targets and an energy research program (IEA, 2007).

From the mid-1990s, the dominant instruments employed to improve energy efficiency in the German manufacturing industries were voluntary agreements. Since its introduction in 2004, however, the emissions trading system has become the most important policy measure in the manufacturing industrial sector, and it has also provided a key incentive to raise energy efficiency (Eichhammer, et al., 2006).

Regarding cross-cutting measures to improve energy efficiency in Germany, the main policy is the Ecological Tax Reform, i.e., the introduction of a so-called Eco Tax on oil, gas and electricity[1]. Additionally, the Renewable Energy Sources Act provides digressive compensation rates for new installations for all renewable energies[2].

The German energy efficiency policies for the manufacturing industries have worked mainly with the following strategies:

- Voluntary agreements: the improvements in the efficiency of on-site electricity generation, particularly combined heat and power (CHP).
- Eco-tax: Germany's red-green coalition government introduced a set of ecotaxes on 1 April 1999 designed to make energy and resource consumption more expensive while lowering the cost of labour. Taxes on petrol and diesel, electricity, heating oil and natural gas had

[1] The tax was introduced in two stages: a first tax increase from 1 April 1999 and a further four-step increase in taxation from 2000 to 2003. There are tax reductions for some consumers, chiefly within the manufacturing industry, agriculture and the railways. The revenue from this tax is used for a reduction of the non-wage labour costs and the promotion of renewable energies (Eichhammer, et al. 2006).

[2] The rates are adapted to the efficiency potential of the different branches. This will provide a strong incentive to reduce costs and increase efficiency (Eichhammer, et al. 2006).

been increased in five stages, and the bulk of the tax revenue generated used to reduce pension insurance contributions.

- Emission trading system means to achieve ecological and economic success. It means assuring the ecological integrity of the instrument, competition neutrality and low transaction costs. In other words, the emission trading system makes use of market-based mechanisms to encourage the reduction of greenhouse gas emissions in a cost-effective and economically-efficient manner, while maintaining the environmental integrity of the system.
- Specific Regulations such as: the Energy Performance of Buildings that seek to promote the energy performance of buildings taking into account outdoor climatic and local conditions as well as indoor climate requirements and cost-effectiveness, and the Minimum Energy Performance Standards for appliances or equipments and mandatory labels that are used to increase the energy efficiency of individual technologies.
- German CHP Law supports of cost efficient technology to reduce CO_2 emissions. This law contains the definition of CHP electricity and heat; support mechanism for high efficiency CHP, and mechanise to supervise reporting of CHP electricity production in CHP plants.
- Renewable Energy Sources Act creates a feed-in tariff system which requires utilities to purchase a predetermined amount of renewable energy at a fixed price. The policy provides economic security for investors and manufacturers and is responsible for the bulk of Germany's dynamic scale-up of renewable electricity capacity and equipment production.
- Grants and loans: the Kreditanstalt für Wiederaufbau (KfW) Umweltprogramm (Environment Program) that provides capital for investment in environmental protection activities and the low-interest loans to SMEs that can be used to supplement the European Recovery Programme's Environment and Energy Saving Program.
- Technology specific rebates are programs used to promote energy management and new energy-efficient technologies.
- Public information and advice: the sub-project under the Initiative Energieeffizienz (Energy Efficiency Initiative) campaign, DENA, the German Energy Agency.

2.2 The Colombian energy efficiency policy

In 1991, with the introduction of the new Constitution, Colombia adopted the principles of sustainable development as a guide to economic development and assigned to municipalities the duty to regulate especially the industry and energy intensive activities. The deregulation of the Colombian electricity system[3] began in the same period, as did the restructuration of the public environmental management system[4]. These elements have characterised the development of energy policies in this country, where the emphasis has

[3]The Colombian electricity industry is characterized by a large hydroelectricity component, close to 70%, and is considered to be one of the most open markets in the developing world, and the market evolution with this model has been satisfactory in terms of investment, competition, efficiency and reduction in electricity losses (Larsen et al., 2004).

[4]The Colombian environmental administration characterizes to be decentralized, democratic, participatory, fiscally solvent, and socially legitimate with measures as a system of pollution taxes, require environmental impact assessments for large construction projects, and institutionalize legal remedies against polluters (Blackman et al., 2006).

been on the formulation of projects and regulations concerning energy efficiency in the manufacturing industrial sector. Moreover, additional instruments for environmental management involve agreements with industry or other relevant organisations. In 1997, the National Environmental Council approved the National Policy of Clean Production. The key objectives of this consensus-based energy policy were to increase the environmental efficiency and quality of energy resources and to develop environmental guides (guias ambientales) detailing options for improving energy efficiency performance in specific sectors. Other strategies used to increase energy efficiency in the manufacturing industries included the establishment of the energy excellence program (Merito URE), the conversion of urban factories from coal or diesel to natural gas and the development of strategies planning for energy efficiency and renewable energy. Currently, the government is developing two legislation projects to improve energy efficiency: Cogeneration Law and the design of the Colombian program of normalisation, accreditation, certification, and labelling of final use of energy equipment.

Hence, Colombian energy policies are based almost entirely on direct regulation. Apart from some small exemptions to VAT taxes for environmental investments, the principal use of economic incentives in energy policies involves the pricing of fuels and agreements with specific manufacturing industrial sector that have high potentials to improve energy efficiency or to carry out changes in technology and renewable energy.

3. Methodology

Changes in energy efficiency were monitored by examining energy use by unit of activity and the application of two indicators of energy efficiency. The first indicator (EI_i) Measures energy use per euro of gross production (equation 1); and the second indicator (CEI_i) Carbon emission intensity the generation of greenhouses gases (in terms of CO_2 emissions) per euro of gross production by each sector i of German and Colombian manufacturing industry (equation 2).

$$EI_i = \frac{E_i}{P_i} \tag{1}$$

$$EI_i = Energy\ intensity\ indicator\ (Mj/€)$$

$$E_i = Energy\ consumption\ of\ the\ textile\ manufacturing\ activity\ i\ (e.g., Mj)$$

$$P_i = Production\ of\ textile\ sector\ i\ (€)$$

$$CEI_i = \frac{CE_i}{P_i} \tag{3}$$

$$CEI_i = CO_2\ emissions\ intensity\ (CO_2\ emissions/€)$$

$$CE_i = CO_2\ emissions\ (tonnes)$$

To identify the factors and variables that influencing energy efficiency in the German and Colombian manufacturing industries, we summarises the opinions and expectations of the main stakeholders (associations and representative firms in Germany and Colombia) through a questionnaire and existing scientific studies.

The questions were designed to identify factors and variables that determine energy efficiency in the manufacturing industries. It included three sections, each with a unique

objective. The first section was designed to establish general information about energy consumption, structure of energy source and energy efficiency.
The second section was designed to assess and rank the importance of different factors and variables in the achievement of improved energy efficiency performance. Questions were asked on issues relating to economic, technical and political factors with their respective variables.
The third section was designed to assess external factors and instruments that would cause or encourage improvements in energy efficiency performance, and what kinds of internal measures or actions would tend to increase energy efficiency performance in the industry.

4. Changes in energy efficiency in German and Colombian manufacturing industries

Energy consumption in the manufacturing industries increased by 2.3% in Germany and 5.5% in Colombia during the sample period (figure 1). The manufacturing industries with the largest increases in energy consumption in this period were paper and tobacco in Germany, and the automotive industry and cement industry in Colombia, whereas the largest decrease in Germany was by the cement industry and in Colombia the machinery industry.

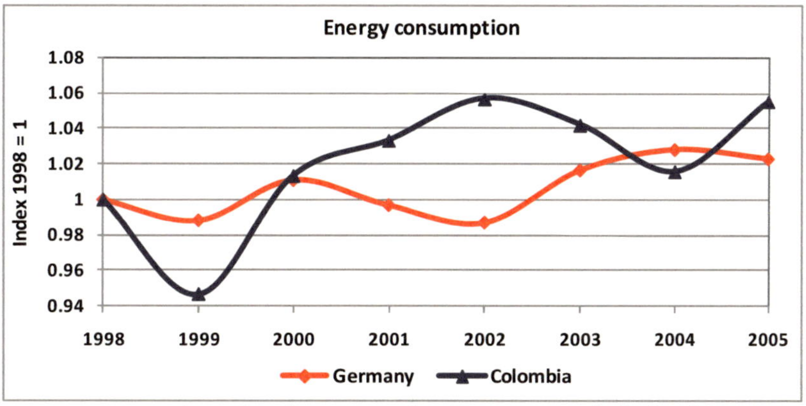

Fig. 1. Energy consumption developments in German and Colombian manufacturing industries

Figure 2 shows developments in average energy intensity for the German and Colombian manufacturing industries between 1998 and 2005. In Germany, this indicator decreased 11% and in the Colombian case decreased 10%. In both countries, several energy intensive sectors have driven the decreases in these indicators for the whole manufacturing sector (in the case of Germany, the chemical industry and basic metal, and in Colombia, basic metal and some sectors of the glass industry).

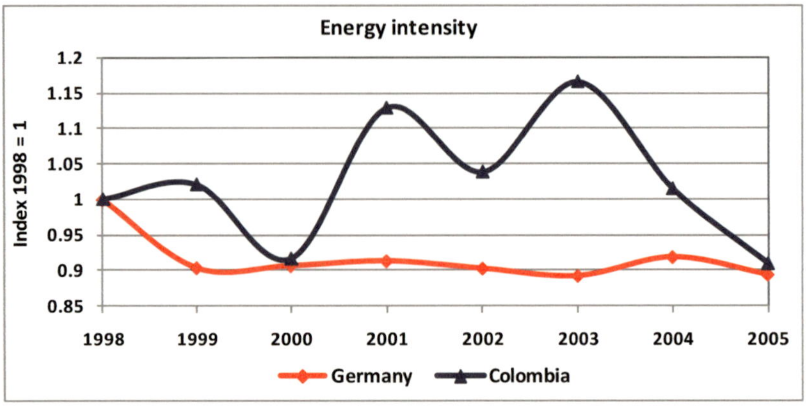

Fig. 2. Energy intensity developments for the German and Colombian manufacturing industries, 1998-2005

The indicator (CEIi) assessed in terms of generation of greenhouse gas emissions, specifically tonnes of CO_2 per gross production. In Germany, the manufacturing industries this indicator decreased 10%. The Colombian manufacturing industries decreased 13% this indicator (see figure 3).

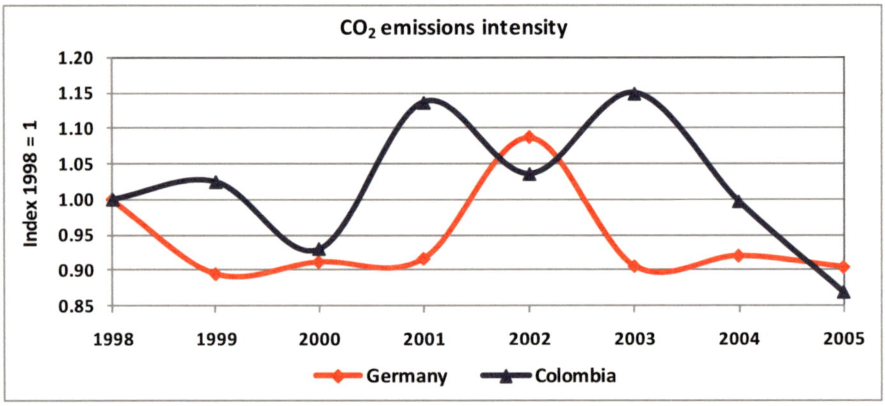

Fig. 3. CO_2 emissions intensity developments for the German and Colombian manufacturing industries, 1998-2005

In Colombia, this indicator in general are still very high in comparison to the German manufacturing industries, and thus there are plenty of opportunities for the Colombian manufacturing industries to further lower this indicator and achieve better and cleaner production figures by improved use of energy resources and a better selection of fuels. By achieving these goals, Colombia will be able to meet international environmental requirements and thus will assure its permanence in the market.

5. Results and discussion

The opinions and expectations of the main stakeholders as primary data are the following: In the German case, two associations and twelve companies, and in the Colombian case, four associations and 26 companies. (see figure 4).

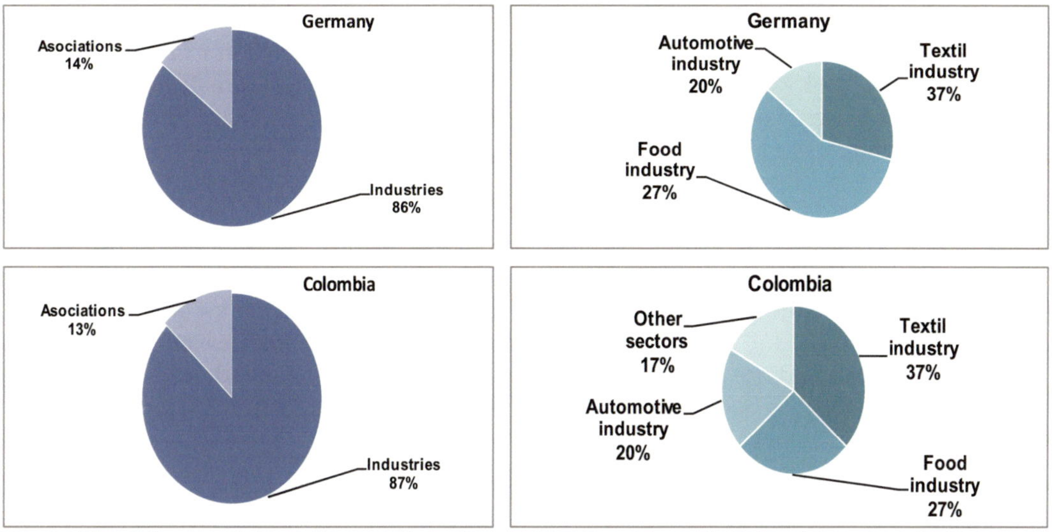

Fig. 4. Breakdown of the primary data from the German and Colombian associations and companies

5.1 Features of energy consumption, energy efficiency and energy source in German and Colombian industries

The results of primary data show that in the German and Colombian cases more than 50% of companies or associations consulted have made studies on energy efficiency and that within of these companies and associations, the majority has analysed and assessed energy efficiency performance and its advantages and disadvantages and included the topic of energy efficiency within their business plans and strategies.

The results also show that the majority of firms and associations know their energy consumption. However, in both countries, the assessment of energy intensity in the companies and associations is a fairly new topic. Moreover, from 2000 to 2008, the assessment of energy consumption and energy intensity has become more prevalent, indicating, possibly, that within the German and Colombian manufacturing industries, the energy topic is becoming more important in the production system and management. This trend would coincide with the increase in certifications of environmental management systems by the countries' in the German case 65% and in the Colombian case 30% by year during this period (ISO, 2007). Hence, energy management is a key program to improve sustainability and environmental performance.

In both countries, the main energy sources for the firms consulted are electricity and natural gas. Energy costs for the firms were between 0.5% and 3% in the German case and between 0.5% and 5% in the Colombian case.

The results in both countries indicate that energy management in the manufacturing industries is important for business strategy and that the quantification and assessment of energy consumption and energy efficiency are input indicators to improve upon in optimisation processes working towards sustainability.

5.2 Factors influencing energy efficiency

In the German case, 43% of firms and associations consider production technology factors very important, and 71% feel that economic and political factors are important in the improvement of energy efficiency performance. In the Colombian case, economic (69%) and production technology factors (62%) are very important factors in achieving improvement of energy efficiency, whereas the political factor is irrelevant (42%) for firms and associations (see figure 5).

These results indicate that in the German case, the firms and associations consider that economic, technical as well as political factors influence energy efficiency, whereas in the Colombian manufacturing industries improvements in energy efficiency are only closely related with economic and production technology factors, mainly because energy efficiency policies are limited and are focalised mainly in support and recommendations of the better technologies.

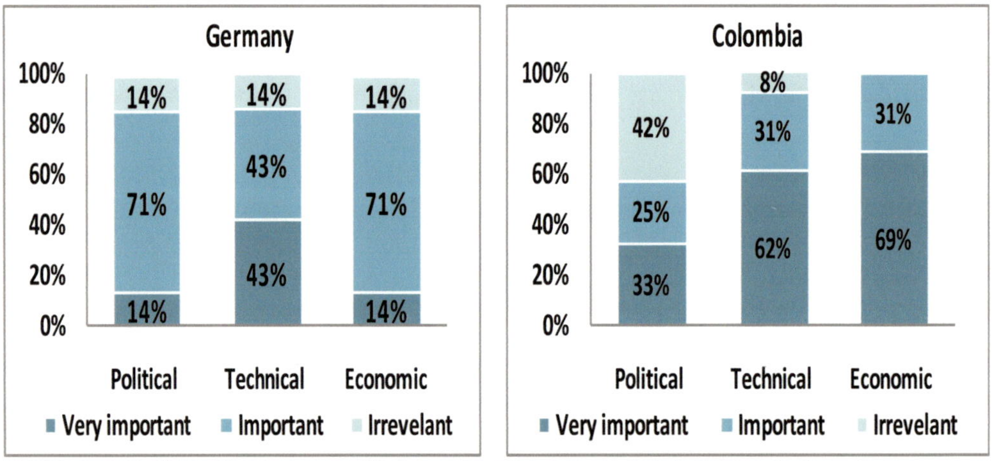

Fig. 5. Factors influencing energy efficiency in German and Colombian industries

- **Variables in economic factors influencing energy efficiency**

Energy consumption in the manufacturing industrial sector is influenced by the behaviour of several economic variables—e.g., high energy prices or constrained energy supply motivate industrial facilities to try to secure the amount of energy required for operations at the lowest possible price (McKane et al., 2008); structural changes in the manufacturing industries cause shifts in final energy use and energy intensities; and the plant capacity utilisation provides an indication of how efficiently plants and equipment are utilised and consequently, could measure the efficiency of energy use.

In the German case, the variables of the economic factor that have the most influence on energy efficiency are improvement in structural operations and maintenance costs and investments in new technologies, equipment or specific activities of energy management investments. Improvements in plant capacity utilisation and levels of production have less importance. On the other hand, in the Colombian case, all variables of the economic factor are important, but the most relevant are improvement in plant capacity utilisation and improvement in levels of production (see figure 6).

These results indicate that manufacturing industries of Germany consider that energy efficiency improvements have higher dependence of investments and production methods, whereas manufacturing industries of Colombia relate energy efficiency improvements with capacity and levels of production. This means that in Germany, improving energy efficiency is important as an investment strategy, whereas in Colombia, energy efficiency is a secondary result from production strategy. This finding concurs with Tholander et al., (2007) who identified the non-priority of energy efficiency investments and lack of access to capital—especially in small and medium enterprises—as main barriers to increased energy efficiency in the manufacturing industries of developing countries in contrast with the situation in developed countries. Moreover, manufacturing industries in developing countries likely prefers traditional investments like expansion of industrial plants or power generation. Furthermore, energy efficiency projects without large capital investments are often perceived as riskier and / or are too small to attract multilateral financial institution lending (UNIDO, 2007).

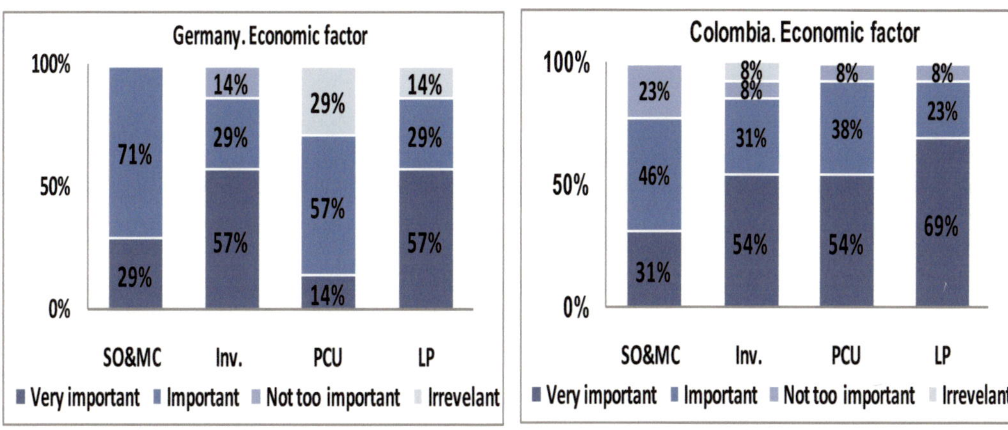

Fig. 6. Variables in the economic factors influencing energy efficiency in German and Colombian industries.

SO&MC: Improvement in the structure of operation and maintenance costs. Inv.: Investments in new technologies, equipments or specific activities of energy management. PCU: Improvement in plant capacity utilisation. LP: Improvement in levels of production.

- **Variables in production technology factor influencing energy efficiency**

The need for improvement of energy efficiency is just one of the drivers for technology development in industry. Moreover, the potential technical energy savings are available based on proven technologies, best practices and use of new energy sources (IEA, 2007).

The manufacturing industries of both countries consider the most important technical variable in improving energy efficiency to be changes in process, operations and machinery. However, for German industries, changes in the structure of energy sources and consumption patterns are also important, while in the Colombian case, in the emphasis is on improved employment behaviour (see figure 7). These results concur with empirical analysis where energy sources emerging as an important variable that influences energy efficiency and in the case of automotive industry and food industry changes of raw materials have been a key variable to improve energy efficiency.

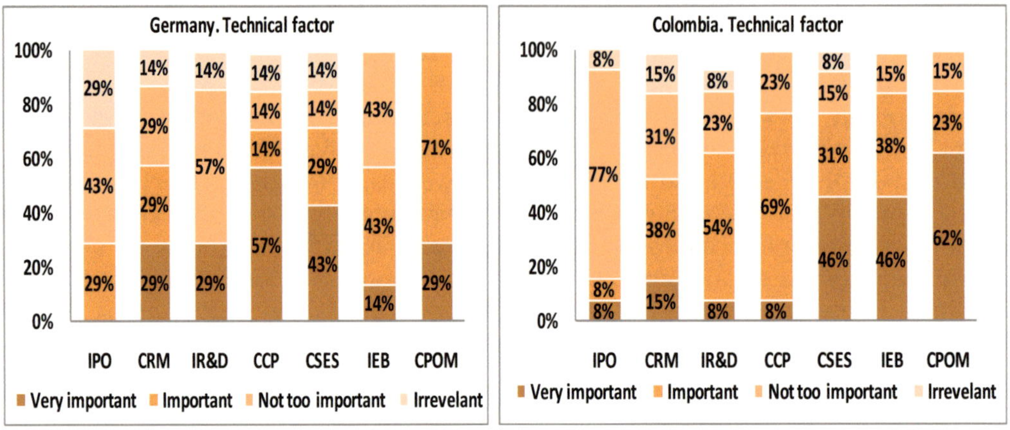

Fig. 7. Variables in the production technology factor influencing energy efficiency in German and Colombian industries.

IPO: Increase processes outsourcing. CRM: Changes of raw materials. IR&D: Increase in the resources of R&D. CCP: Changes of consumption patterns. CSES: Changes in the structure of energy sources. IEB: Improvements in employment behaviour. CPOM: Changes in the process, operations and machinery.

These results show that the manufacturing industries of both countries feel that the best way to improve energy efficiency is by changes in process, operations and machinery (Germany 71% and Colombia 62%) generally these processes in the organizations begin with an internal analysis of the production process and machinery to determine opportunities to decrease energy consumption and increase energy efficiency. Moreover, in the Colombian case, it's also important the analysis of employment behaviour because behaviour change erodes the energy savings due to the technical energy efficiency improvements, especially in developing countries (IEA, 2005).

Hence, the results confirm that Germany has achieved important developments in energy efficient-technology and significant improvement in energy efficiency performance in the manufacturing industries. According to the Federal Ministry of Economics and Technology, Germany in recent years has achieved a decrease in its energy consumption even though the gross domestic product has more than doubled and German researchers and companies have submitted many global patent applications in the development of energy efficient industrial cross application technologies.

- **Variables in political factors influencing energy efficiency**

Market forces and other factors determine energy efficiency in the manufacturing industries. However, these factors can be influenced by an effective energy policy that encourages cost effective energy efficiency through the application of different types of policy instruments that include information, regulation and economic instruments.

Figure 8 shows the results of variables in the political factors affecting energy efficiency in German and Colombian industries. In the German case, the most important variables of the political factor are to encourage the application of energy management in the organizations, mandatory standards (such as the efficiency of electric motors and the efficiency of industrial boilers), and soft loans—especially for cogeneration (CHP). These results concur with Eichhammer, et al. (2006), who showed that only some measures are seen as a high-impact (the first voluntary agreement with German industry from 1995 and the second financial measures (CHP Act, KfW Umweltprogramm)), whereas the impact of the Ecological Tax Reform has been estimated as medium, and other measures have been assessed as low-impact.

However, according to studies of Ecofis et al., (2206) voluntary agreements to save energy are adequate in these circumstances when dealing with a small number of actors with which you need to negotiate or a strongly organized sector and / or when there is much relatively cheap energy saving potential. The characteristics that could determine the success of this instrument are the following: the target group motivated to participate, there are penalties in case of non- compliance, there is a good monitoring system, and adequate supporting instruments such as audits, energy monitoring systems, financial incentives and demonstrations projects.

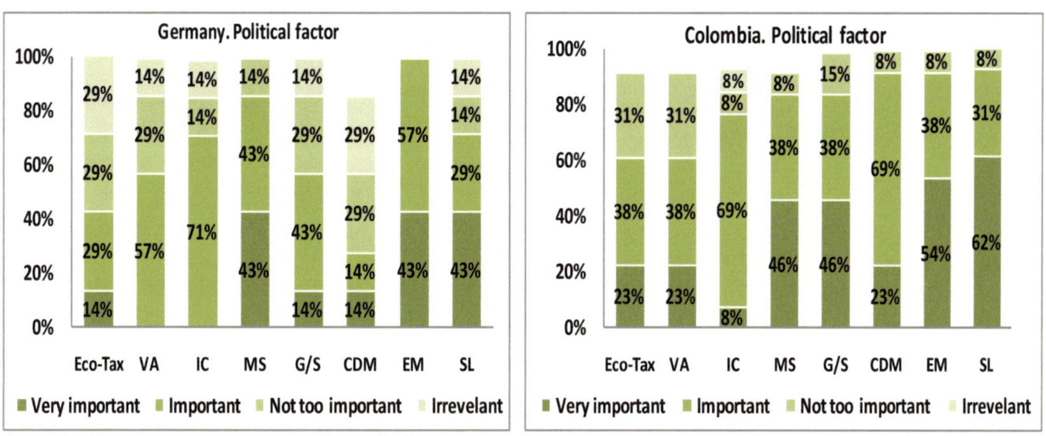

Fig. 8. Variable in the political factors influencing energy efficiency in German and Colombian industries

Eco-tax: Eco-tax.VA: Voluntary audits. IC: Information campaigns. MS: Mandatory standards (the efficiency of electric motors and the efficiency of industrial boilers). G/S: Grants / subsidies. CDM: Emission trading / Clean Development Mechanism. EM: to encourage the application of energy management SL: Soft Loans for Energy Efficiency, Renewable energy and CHP.

In the Colombian firms, the most important variables are soft loans (for Energy Efficiency, Renewable energy and cogeneration (CHP)), to encourage energy management and the emissions trading / Clean Development Mechanism—indicating that in this country, a barrier to improved energy efficiency is the limited amount of resources available to change technology and to achieve improved energy efficiency, a conclusion which concurs with the studies of Kant, 1995; Tanaka, 2008 and Gillingham et al., 2009.

5.3 Instruments influence interest to improve energy efficiency performance

Figure 9 shows that instruments and measures would cause or encourage the German and Colombian manufacturing industries to improve energy efficiency performance. In both countries, the main instruments are changes in upstream sector (energy prices) and institutional regulations, whereas labelling to have a lower impact.

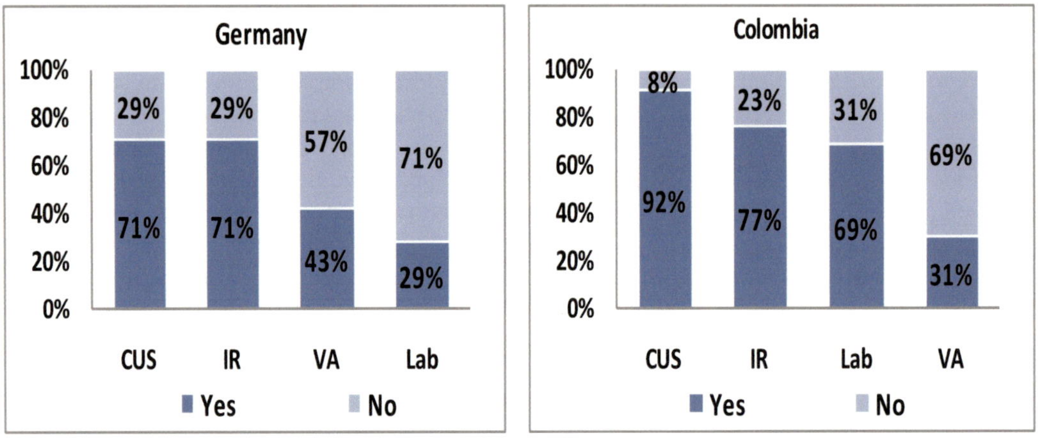

Fig. 9. Percentage of respondents who felt that specific measures and instruments could improve energy efficiency performance

CUS: Changes in upstream sector (energy prices). IR: Institutional regulations (Regulatory standards, - Fiscal policy, State aid for R&D). VA: Voluntary agreements. Lab: Labelling (e.g. industrial motors, EMAS, ISO 14001).

The results are clear in the German case, where a series of energy-conservation instruments have been implemented to include: the replacement of traditional gas- or oil-fired boilers with condensing gas-fired boilers, the gradual replacement of traditional fuels with more expensive bio-fuel, and the consecutive emergence of integrated gasification combined cycle (CGC) and combined heat and power (CHP) systems. As a result, the energy intensity of Germany has decreased 20% from 1990 to 2003, with an annual decrease rate of 1.75%. Moreover, during the last decade, the energy policy of Germany has been strongly influenced by environmental issues, and the German government has consecutively introduced various acts related to renewable energy and energy efficiency. During 1999, to stimulate energy conservation, energy efficiency, and the application of renewable energy technologies, the German government introduced the Eco-tax, which subsequently became the Renewable Energy Act, which targets a short-term goal of doubling renewable power generation by 2010, together with an intermediate-term goal of increasing renewable power generation capacity to 20% of total power generation capacity by 2020 (Blesl et al., 2007).

5.4 Internal measures and actions the manufacturing industries would consider to increase energy efficiency performance

Figure 10 shows the kinds of internal measures and actions the manufacturing industries would consider to increase energy efficiency performance. In the German case, the most important internal measures in order of importance are energy management systems, energy efficiency investments, and changes in machinery and equipment. In the Colombian case, the most important internal measures in order of importance are energy efficiency investments, changes in machinery and equipment, and optimisation of production capacity and production level.

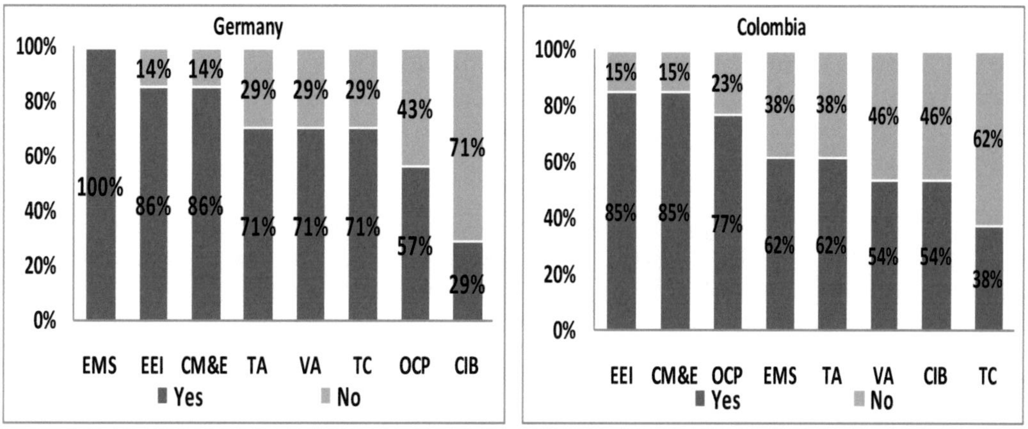

Fig. 10. Kinds of internal measures and actions the manufacturing industries would consider to increase energy efficiency performance

EMS: Energy management systems. EEI: Energy efficiency investment (e.g. changes in machinery, equipments and technology). CM&E: Changes in machinery and equipment. TA: Training activities. VA: Voluntary audit. TC: Major product/process related technological changes, whether or not introduced as part of public/private national and the R&D programmes. OCP: Optimization of production capacity and production level. CIB: Conversion of industrial business (in terms of both products and processes).

These results show that in both countries, the manufacturing industrial sector has an interest in increasing their investments to improve energy efficiency through changes in machinery and equipment—demonstrating that the manufacturing industrial sector considers improvements in energy efficiency to be closely related with technological change. This result coincides with opportunities to improve industrial energy efficiency through new technologies such as the use of high-efficiency motor-driven systems, the optimisation of compressed air systems and the potential that exists based on currently available improvements. In fact, the possibility of implementing new and emerging technologies with potential savings of as much as 35 percent in energy costs is creating entirely new lines of business (IAC, 2007).

Finally, the results of this study suggest that policy strategies in the manufacturing industries have to utilise legal and fiscal instruments to generate supporting framework conditions as well as targeted programs in the fields of R&D, technological change, market transformation, information, education, dissemination of best practice, etc. Moreover, policy will always have to live with unavoidably sub-optimal solutions, while growing knowledge and changing frameworks will constantly impose the need to search for better solutions and new opportunities. In this context, energy policy strategies represent not only (static) problems of policy choice but—above all—dynamic search and learning processes aimed at designing effective policy measures.

6. Recommendations for the formulation of energy-efficiency policies in the Colombian manufacturing industrial sector

According to our results and the literature, it is important that there be a formulation of an adequate package of policies and measures that are addressed to guarantee effective and efficient impact to improve energy-efficiency performance and reducing greenhouse emissions in the Colombian manufacturing industries. The following strategies and instruments in policy settings are recommended in order to achieve improvements in energy efficiency in a cost-effective manner:

a. Policy support. Policy support should aim at making energy efficiency easy ("Make it easy!"), realisable ("Make it possible!"), and beneficial ("Make it rewarding!") for stakeholders, thereby contributing to the development of the market for energy-efficient technologies and services. Due to the implementation of the support programmes, it also becomes clear that energy efficiency is politically intended and crucial ("Make it a policy!"). A pre-planned, target-group-specific, differentiated mix of policy instruments and measures is necessary, with integrated measures that are directly addressed to stakeholders. In such a way, the specific situations, incentives, barriers and obstacles of different stakeholders should be addressed by specific policy mixes (Thomas and Irrek, 2007).

b. Integral approach. The most effective way to improve industrial energy efficiency is through an integrated approach, where a number of policies and programmes are combined to create a strong overall industrial energy-efficiency policy that addresses a variety of needs in Colombian manufacturing sectors. There should thus be an adoption of a policy of energy-efficiency sector targets and related programmes in which individual manufacturing industrial sectors committed to specific improvements in energy intensity over a given time period in exchange for governmental support in the form of financial incentives, information programmes, demonstration programmes, and training programmes, significant energy savings could be realised.

c. Energy efficiency strategies. National energy efficiency strategies in Colombia could accelerate the implementation of energy efficiency in the manufacturing industries. National energy-efficiency strategies should be useful because during their development, implementation and evaluation, they can help to achieve the following: make the vision for energy efficiency explicit; focus attention on the important issues; identify gaps in current work programmes; identify necessary tasks and resources and allocate implementation and monitoring responsibility.

d. Energy data. The Colombian government through the statistical office and energy agency (UPME) must improve the availability of high-quality energy efficiency data because without accurate energy time series data, it is difficult to target and develop appropriate energy efficiency policies in the manufacturing industries. Moreover, for developing sectoral energy efficiency benchmarks and best practices, action plans should: assess energy consumption by end-use in manufacturing industrial sector; identify the economy's energy-saving potentials and establish objectives and adequate methods for evaluating the success of the plan.

e. Mandatory standards. For the Colombian manufacturing industrial sector, the most important technical variable to improve energy efficiency is change in processes, operations, machinery and equipment. For this reason, the Colombian government should consider adopting mandatory minimum energy performance standards for machinery and equipment (e.g., the efficiency of industrial motors and the efficiency of industrial boilers) in line with international best practices. Moreover, it should examine barriers to the optimisation of energy efficiency through technology systems and design and implement comprehensive policy portfolios aimed at overcoming such barriers.

f. Energy management. Among Colombian firms, one of the most important political variables is the encouragement of the application of energy management[5]. The Colombian government should thus consider providing effective assistance in the development of energy management (EM) capability through the development and maintenance of EM tools, training, certification and quality assurance. Moreover, it should encourage or require major industrial energy users to implement comprehensive energy management procedures and practices that could include, according to IEA, 2008:

- The development and adoption of a formal energy management policy. The process and implementation of this policy should be reported and overseen at the company board level and reported in company reports. Within this policy, companies would need to demonstrate that effective organisational structures have been put in place to ensure the following: that decisions regarding the procurement of energy-using equipment are taken with the full knowledge of the equipment's expected life-cycle costs and that procurement managers have an effective incentive to minimise the life-cycle costs of their acquisitions.
- The appointment of full-time qualified energy managers at both the enterprise- and plant-specific levels as appropriate.
- The establishment of a scheme to measure, monitor, evaluate and report industrial energy consumption and efficiency at the individual company sector and national levels. As a part of this effort, appropriate energy performance benchmarks should be developed, monitored and reported at levels deemed suitable for each sector.

g. Small and Medium-sized Enterprises (SMEs). The size of company variable was significant for Colombian industry. The Colombian government should thus consider

[5] There are significant cost-effective energy savings to be realised in industry through the more widespread adoption of best practices in energy management (EM). EM addresses the way in which an industrial plant or facility is managed to identify and exploit cost-effective energy savings opportunities (IEA, 2008).

developing and implementing a package of policies and measures to promote energy efficiency among SMEs. This package should include: a system for ensuring that energy audits, carried out by qualified engineers, are widely promoted and easily accessible for all SMEs; the provision of high-quality and relevant information on energy-efficiency best practices; the provision of energy performance benchmarking information that ideally would be structured to allow international and national economy comparisons; and appropriate incentives to adopt capital acquisition and procurement procedures with the lowest life-cycle costs.

h. Investments. For the Colombian manufacturing industrial sector, the results indicate that energy efficiency investments are a key variable to improve energy efficiency. However, among the many impediments to the adoption of cost-effective energy efficiency investments is the "finance barrier" (Tholander et al., 2007 and IEA, 2008). The Colombian government should facilitate the manufacturing industrial sector's and stakeholders' involvement in energy efficiency investments by: I) adopting and publicising to the manufacturing industrial sector a common energy-efficiency savings verification and measurement protocol in order to reduce existing uncertainties in quantifying the benefits of energy efficiency investments and stimulate increased private sector involvement; II) reviewing their current subsidies and fiscal incentive programmes to create more favourable grounds for private energy-efficiency investments; III) collaborating with the private financial sector to establish public-private tools to facilitate energy-efficiency financing; IV) promoting risk-mitigation instruments such as securitisation or public-private partnerships; V) putting in place institutional frameworks to ensure regular co-operation and exchanges on energy efficiency issues between the public sector and financial institutions and VI) design an energy tax programme to provide an incentive to industry to improve energy management at firms' facilities through both behavioural changes and investments in energy-efficient equipment.

i. Taxes and tariff structure. This study demonstrated that energy costs and taxes are important for improving energy efficiency. The Colombian government should design a package of taxes and a tariff structure that include the following: I) the reduction of subsidies or using energy to balance the effect of subsidies, providing the energy consumer with a more realistic indication of the actual costs associated with certain forms of energy; II) the use of taxes to more accurately reflect the environmental costs, or "externalities", associated with energy consumption; III) the imposition of taxes and fees associated with energy use resulting from energy consumption on users with the goal of creating incentives to reduce wasteful energy consumption practices or creating public programmes and funds for encouraging energy efficiency and IV) having the price system ensure that all individual agents are confronted with the full costs that their decisions impose on others; this means addressing externalities and market failures through a greater use of taxes, charges and tradable permits and correcting policy failures through reforms of support programmes that are environmentally harmful and economically inefficient and have undesirable social effects.

j. Control, monitoring and evaluation. Developing effective energy-efficiency policies requires a good understanding of how energy is used as well as the various factors that drive or restrain demand. Such an understanding requires accurate data on energy end-use and the associated activities. The Colombian government should thus ensure that instruments of energy efficiency policies are adequately monitored, enforced and evaluated so as to ensure maximum compliance and that their energy-efficiency policies are supported by adequate end-use information by substantially increasing their effort to collect energy end-use data across all sectors and relating to all energy types.

k. Technology transfer and cooperation. In the Colombian manufacturing industrial sector, this analysis demonstrated that the technology level is still moderate and that this technical factor is a key strategy to improve energy efficiency. The Colombian government should thus promote technology transfer through an appropriate enabling framework in order to enhance international cooperation for the scaling up of sustainable energy solutions. The transfer of technology requires a careful balancing act that includes both fair treatment for innovators and energy policies that stimulate global diffusion of energy technology to address energy efficiency.

7. Conclusions

In this chapter analysed the energy efficiency in German and Colombian manufacturing industries in the time period 1998-2005 using economic indicators. We found that the industrial sectors of both countries during the sample period increased their energy consumption by 2.3% in Germany and 5.5% in Colombia and also decreased the energy intensity (11% and 10% respectively). Therefore, German and Colombian manufacturing industries improved energy efficiency and decreased CO_2 emissions demonstrating that the trend of manufacturing industry is "make more with less energy consumption and clean production".

Based on the primary data from German and Colombian industrial associations and representative firms in each country, the economic, technical and political factors were studied with respect to impact on energy efficiency. The results in both countries indicate that energy management for the manufacturing industrial sector is important within business strategy and that the quantification and assessment of energy consumption and energy efficiency are input indicators to be used in improvement and optimisation processes within sustainability development.

The results also show that in German industry, economic, technical and political factors influence energy efficiency, whereas in the Colombian case, improvements in energy efficiency are closely related with economical and production technology factors.

In the German case, the results showed the following: (I) the variables in the economic factor with the most influence on energy efficiency are the structural operations and maintenance costs and investments, whereas plant capacity utilisation and levels of production have lower importance. (II) The most important technical variables to improve energy efficiency are changes in the processes, operations and machinery, changes in the structure of energy sources, and changes of consumption patterns. (III) The most important variables in the political factor are to encourage the application of energy management, mandatory standards (such as the efficiency of electric motors and the efficiency of industrial boilers), and soft loans especially for cogeneration (CHP). (IV) The most important internal measures to improve energy efficiency are energy management systems, energy efficiency investment, and changes in machinery and equipment.

In the Colombian case, the results showed the following: (I) All variables for the economic factor are important, but the most relevant are plant capacity utilisation and levels of production. (II) The most important technical variables to improve energy efficiency are changes in the processes, operations and machinery, and improvements in employment behaviour. (III) The most important variables of the political factor are soft loans (for Energy Efficiency, Renewable energy and cogeneration (CHP)), to encourage the application of energy management and emissions trading / Clean Development Mechanism. (IV) The most important internal measures for increasing energy efficiency are energy efficiency investments, changes in machinery and equipment and optimisation of production capacity and production level.

Moreover, the results suggest that policy strategies in the Colombian manufacturing industrial sector have to combine the following strategies: integral approach, energy data, mandatory standards, energy management, the promotion of energy efficiency in small and medium-sized enterprises, investments, a tax program, an adequate tariff structure, control and evaluation, technology transfer and cooperation.

Acknowledgments

The author would like to thank Professor Dr Werner Bönte, Dr Wolfang Irrek and Dr Alexander Cotte Poveda for the helpful suggestions and comments. The author is grateful for the support provided by the Wuppertal Institute, Deutscher Akademischer Austausch Dients and the University of La Salle. Any remaining errors are the responsibility of the author.

8. References

Blesl M., Das A., Fahl U., Remme U. (2007). Role of energy efficiency standards in reducing CO2 emissions in Germany: An assessment with TIMES. Energy Policy 35, 772-785.

Blackman A, Morgenstern R., Montealegre L., Murcia L., and García J. (2006). Review of the efficiency and effectiveness of Colombia's environmental policies. An RFF Report.

Kant A. (1995). Strategies and Instruments to promote energy efficiency in developing countries. Project working paper 5. Effectiveness of industrial energy conservation programmes in IEA countries ECN-C-94-113.

ECOFYS, Wuppertal Institut, Lund University. (2006). Guidelines for the monitoring, evaluation and design of energy efficiency policies. How policy theory can guide monitoring and evaluation efforts and support the design of SMART policies. www.aid-ee.org

Eichhammer, W., Schlomann, B., Kling N. (2006). Energy Efficiency Policies and Measures in Germany 2006. Monitoring of Energy Efficiency in EU 15 and Norway (ODYSSEE-MURE). Fraunhofer Institute for Systems and Innovation Research (Fraunhofer ISI).

Gillingham K., Newell R., Palmer K. (2009). Energy efficiency economics and policy. Working Paper 15031. http://www.nber.org/papers/w15031

International Energy Agency (IEA). (2008). Energy efficiency policy recommendations. In support of the G8 Plan of Action. http://www.iea.org/G8/2008/G8_EE_recommendations.pdf

International Standard Organisation (ISO). (2007). The ISO Survey of Certifications 2006. www.iso.org

Inter Academy Council (IAC). (2007). Lighting the way. Toward a sustainable energy future. www.interacademycouncil.net

International Energy Agency (IEA). (2005). The experience with energy efficiency policies and programmes in IEA countries. Learning from the critics. IEA Information paper.

International Energy Agency (IEA). (2007). Tracking Industrial Energy Efficiency and CO2 Emissions. In support of the G8 Plan of Action. Energy Indicators.

Larsen E., Dyner. I, Bedoya L., Franco C. (2004). Lessons from deregulation in Colombia: successes, failures and the way ahead. Energy policy 32, 1767-1780.

McKane A., Price L., Rue S. (2008). Policies for Promoting Industrial Energy Efficiency in Developing Countries and Transition Economies. United Nations Industrial Development Organization.

Tanaka K. (2008). Assessment of energy efficiency performance measures in industry and their application for policy. Energy policy (2008), doi:10.1016/j.enpol.2008.03.032.

Thomas S., Irrek W. (2007). Wie 20 Prozent Endenergieeinsparung möglich werden können. Worschläge des Wuppertal Instituts zum deutschen Energieeffizinez-Aktionsplan und zu Maßnahmen im Industriebereich. VIK Mitteilungen 3/07, 16-18.

Thollander P., Danestig M., Rohdin P. (2007). Energy policies for increased industrial energy efficiency: Evaluation of a local energy programme for manufacturing SMEs. Energy policy 35, 5774–5783.

United Nations Industrial Development Organization (UNIDO). (2007). Policies for promoting industrial energy efficiency in developing countries and transition economies. Commission for Sustainable Development (CSD-15).

Wuppertal Institute, 2008. Greenhouse Gas Mitigation in Industry in Developing Countries. Final Report. On behalf of the Deutsche Gesellschaft für Technishe Zusammenarbeit (GTZ). http://www.wupperinst.org/en/projects/proj/index.

5

Oxyfuel combustion in the steel industry: energy efficiency and decrease of CO_2 emissions

Joachim von Schéele
The Linde Group
Germany

1. Introduction

The use of oxygen technologies within the steel industry has become increasingly important. During the last decades increased throughput capacity and lowered average cost have been the driving forces, however, today the positive impact on energy savings and reduced emissions have come into the focal point, a fact that seems to be even further pronounced in the future. This chapter describes how the oxygen technologies contribute to increased energy efficiency in the melting and heating processes, how it reduces the fuel consumption and CO_2 emission, and how in-plant generated low calorific gases can be effectively used to further improve the overall energy efficiency of a steel production plant, reduce costs and environmental impact.

The main production routes for steel are the integrated steel mill and the mini-mill. The integrated steel mill uses iron ore as main source for iron, and includes processes like ore sintering, coke-making, blast furnace iron-making and basic oxygen steel-making. The main piece of equipment at a mini-mill is the electric arc furnace where steel scrap, its main raw material, is melted. Both routes include subsequent casting and downstream heating and rolling (or forging) operations.

Dependent on production route and status, a steel mill need 700 to 4,000 kWh to produce 1 tonne of finished product. This corresponds to a CO_2 emission of about 0.35 to 2.2 tonne per tonne of steel produced. However, there are great opportunities to increase the efficiency, using oxygen technologies make a substantial positive impact. Relating to how the oxygen is introduced, we basically distinguish between injection of oxygen (normally through a lance) and oxyfuel combustion (applying a burner), however, the end result is the same: oxyfuel combustion. The main processes where oxygen technologies can be applied are: electric arc furnace for scrap melting, blast furnace iron-making, preheating of different vessels (ladles, etc.), and in the downstream reheating and heat treatment.

It is a well-known fact that only three things are needed to start and maintain combustion: oxygen, fuel, and sufficient energy for ignition. The combustion process itself would be most efficient if fuel and oxygen can meet without any restrictions. However, in practice it is not simply a question of efficient combustion, the heat transfer efficiency is also extremely important. Nevertheless, it has been clearly demonstrated that if oxygen (and not air) is used to combust a fuel, all the heat transfer mechanisms (convection, conduction and radiation) can be promoted at the same time. Air contains 21% oxygen and 79% ballast. In a combustion process, this ballast, practically all nitrogen, has to be heated, without taking part in the process. By using oxygen instead of air we get the beneficial oxyfuel combustion.

New demands and challenges from the industry have been met by a continuous development work. As a result, in parallel to the conventional oxyfuel - for example widely used to boost melting in electric arc furnaces - there are today established very interesting technologies. Among those, the most important ones seem to be flameless combustion and direct flame impingement. These new technologies not only fulfil the existing needs with astonishing results, they also open up for completely new areas of application.

Flameless oxyfuel is today applied in drying and preheating of ladles and converters, for heating in reheat furnaces and annealing lines, and for melting when avoiding oxidation. It provides excellent temperature uniformity and reduced NO_X emissions. Additionally, it can be applied in, for example, preheating of air in the blast furnace hot stoves.

The use of direct flame impingement has so far been limited to boosting of strip annealing and galvanizing lines, but its opportunities are almost uncountable. For example, there are ideas about applying this technology to substantially shorten process routes by omitting process steps, or using it in the iron-making step.

In reheating, today's best air-fuel solutions need at least 1.3 GJ (360 kWh) for heating a tonne of steel to the right temperature for rolling or forging; employing oxyfuel the comparable figure is below 1 GJ, a saving of 25%. For continuous heating operations it is also possible to economically operate the furnace at a higher temperature at the entry (loading) side of the furnace. This will even further increase the possible throughput in any furnace unit. Oxyfuel combustion allows all installation pipes and flow trains to be compact without any need for recuperative or regenerative heat recovery solutions. Combustion air-blowers and related low frequency noise problems are avoided.

Oxyfuel solutions deliver a unique combination of advantages in reheat and annealing. Thanks to improved thermal efficiency (about 80% compared with 40-60% for air-fuel), the heating rate and productivity are increased and less fuel is required to heat the product to the desired temperature, at the same time saving on CO_2 and NO_X emissions. In summary the results include:

- Throughput capacity increase of up to 50%
- Fuel savings of up to 50%
- Reduction of CO_2 emissions by up to 50%
- Reduction of NO_X emissions
- Reduction of scaling losses (improving the material yield)

Compared with conventional oxyfuel, flameless oxyfuel provides even higher production rates, excellent temperature uniformity and very low NO_X emissions. Since its commercial introduction in 2003, the leading supplier has made more than 30 installations of the flameless oxyfuel technology, some using a low calorific fuel.

This chapter describes the state-of-the-art of oxygen technologies, including results from installations in the steel industry, and discusses their future very interesting possibilities to make the steel production more effective. Oxyfuel combustion has begun to make the steel industry more energy efficiency, but more can be done and, moreover, those technologies can be employed also in other branches of the industry, there as well making improvements of 20-50%.

2. Oxyfuel combustion technologies

Oxyfuel combustion refers to the use of pure, that is industrial grade, oxygen instead of air for combustion of fossil fuels. Oxyfuel technology offers a number of advantages over air-fuel combustion. In air-fuel combustion the burner flame contains nitrogen from the combustion air. A significant amount of the fuel energy is used to heat up this nitrogen. The hot nitrogen leaves through the stack, creating energy losses. When avoiding the nitrogen ballast, by the use of industrial grade oxygen, then not only is the combustion itself more efficient but also the heat transfer. Oxyfuel combustion influences the combustion process in a number of ways. The first obvious result is the increase in thermal efficiency due to the reduced exhaust gas volume, a result that is fundamental and valid for all types of oxyfuel burners. In combustion gases, heat radiation is mainly from CO_2 and H_2O molecules. As there is no, or very low, nitrogen content in an oxyfuel furnace atmosphere, the concentration of highly radiating CO_2 and H_2O will be very high, a fact which considerably increases heat transfer by gas radiation. A striking feature of oxyfuel combustion is the very high thermal efficiency even at high flue-gas temperatures and no preheating of fuel or oxygen.

Fig. 1. An ingot for bearing steel production is lifted out of a soaking pit furnace at Ascométal in France. The furnace is fired with flameless oxyfuel, heating the ingots uniformly to over 1200°C.

In addition to using a burner for the combustion, which normally is operated at stoichiometric conditions, two other technologies should be mentioned: lancing, and post-combustion. Lancing refers to injecting oxygen, sometimes at very high velocities into furnace free-space or a melt. It is done to intensify the air-fuel combustion, either to combust for example carbon into CO, or achieve a complete combustion of a fuel into products like CO_2 and H_2O. Typically it is employed an Electric Arc Furnace (EAF) for scrap melting, but it could also be like in the case of the REBOX® HLL technology to improve reheating.
Post-combustion does in most cases in this context refer to the reaction $CO+\frac{1}{2}O_2=CO_2$, which is strongly exothermic; the released energy is typically used to improve melting. Generally speaking, the prerequisites for a beneficial post-combustion are CO generation,

oxygen available, and a high heat transfer. For example, charging coal with the scrap in an EAF so that it dissolves into the hot heel and blowing oxygen into the hot heel at simultaneous over-stoichiometric operation of the oxyfuel burners when there is scrap in the furnace, provides such wanted conditions. Post-combustion at flat bath operation, on the other hand, normally provides too low heat transfer efficiency.

2.1. Flameless oxyfuel combustion

Some very interesting technologies have emerged in parallel with conventional oxyfuel, which is widely used to boost melting in electric arc furnaces. The most important ones are flameless combustion and Direct Flame Impingement (DFI). These new technologies not only fulfil existing needs with astonishing results, they also open up completely new areas of application. The flameless combustion creates a huge practically invisible oxyfuel flame whereas the DFI technology uses small, well-defined sharp flames.

Increasingly stricter legislation on emissions led to the development of flameless oxyfuel, which was introduced for the first time in 2003 in continuous furnaces for strip annealing and slabs reheating, both at the stainless steel producer Outokumpu. The expression 'flameless combustion' communicates the visual aspect of the combustion type, that is, the flame is no longer seen or easily detected by the human eye. Another description might be that combustion is 'extended' in time and space – it is spread out in large volumes, and this is why it is sometimes referred to as 'volume combustion'. Such a flame has a uniform and lower temperature, yet containing same amount of energy.

In flameless oxyfuel the mixture of fuel and oxidant reacts uniformly through flame volume, with the rate controlled by partial pressures of reactants and their temperature. The flameless oxyfuel burners effectively disperse the combustion gases throughout the furnace, ensuring more effective and uniform heating of the material even with a limited number of burners installed. The lower flame temperature is substantially reducing the NO_X formation. Low NO_X emission is also important from a global warming perspective; NO_2 has a so-called Global Warming Potential that is almost 300 times that of CO_2.

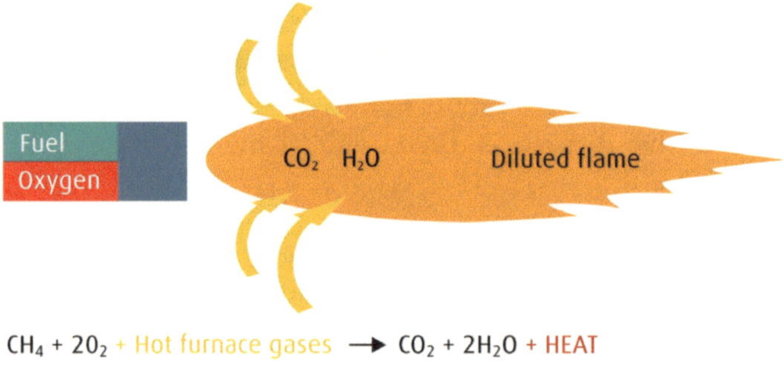

Fig. 2. The principle way of creating flameless combustion; the flame is diluted by the hot furnace gases. This reduces the flame temperature to avoid creation of thermal NO_X and to achieve a more homogenous heating of the steel.

Compared with conventional oxyfuel, flameless oxyfuel provides even higher production rates, excellent temperature uniformity and very low NO_X emissions. The first installations of this innovative flameless oxyfuel technology were made by Linde. Since 2003 over 30 installations of this technology have been made at more than a dozen sites, some even using a low calorific fuel. There seems to be an increasing need to combust low calorific fuels; for fuels containing below 2 kWh/m^3, use of oxygen is an absolute requirement for flame temperature and stability. At integrated steel mills use of blast furnace top gas (<1 kWh/m^3), alone or in combination with other external or internal fuels, is becoming increasingly important. Flameless oxyfuel can be successfully employed here.
The first installations of flameless oxyfuel took place in reheating and annealing, but it was quickly adopted also for preheating of ladles and converters. The next area to be exploited, with substantial positive impact, is the blast furnace hot stoves.

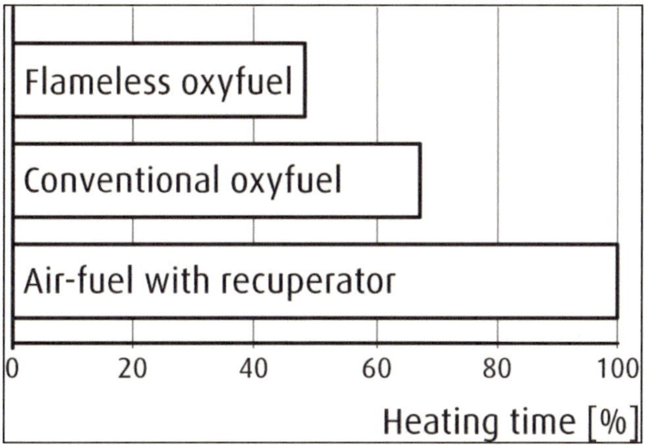

Fig. 3. A comparison of the results from installations at Ovako's Hofors Works, Sweden using different combustion technologies. When conventional oxyfuel was installed the heating time decreased by 30%, but with flameless oxyfuel it was possible to run a heating time half of the original one with air-fuel. It should be noted that the power has not been increased, but decreased.

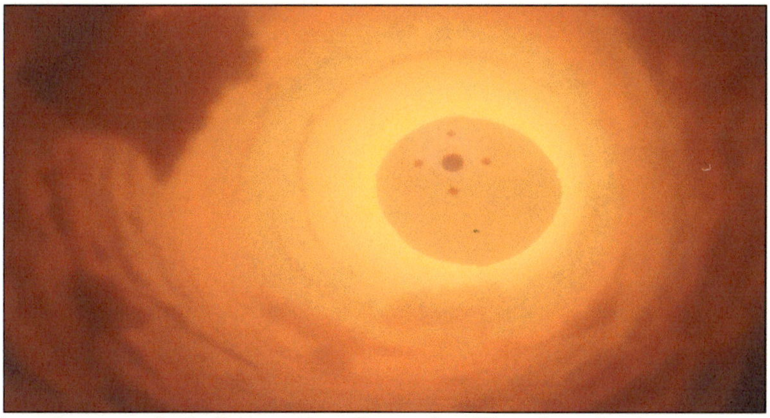

Fig. 4. A flameless oxyfuel flame; the flame is diluted and almost transparent. The combustion in this photograph is using a low calorific fuel.

2.2. Direct Flame Impingement

It is also possible to fire with oxyfuel flames directly onto a material. This is what we call DFI, Direct Flame Impingement. DFI Oxyfuel is a fascinating compact high heat transfer technology, which since 2002 provides enhanced operation in strip processing lines, for example at galvanizing. It is patented by Linde. So far the use of DFI Oxyfuel has been to boost strip annealing and hot dip metal coating lines. Use of DFI Oxyfuel reduces the specific fuel consumption while delivering a powerful 30% capacity increase, or more. Installations are found at Outokumpu's Nyby Works in Sweden and ThyssenKrupp at Finnentrop and Bruckhausen in Germany. The latest installation is in a continuous annealing line at POSCO in Pohang, South Korea. Due to DFI's effective pre-cleaning properties there are also benefits relating to surface appearance and improved adhesion of metal coatings.

The main benefits of DFI Oxyfuel are:

- Significantly higher heat transfer resulting in increased production capacity
- Lower fuel consumption
- Ability to use high power input in limited furnace volume
- Compact and powerful unit for easy retrofit
- Heating and cleaning in one operation
- Option to modify surface conditions

Tests have verified the higher level of local heat flux for the DFI Oxyfuel technology. In general, the use of oxyfuel combustion substantially increases the thermal efficiency of a furnace. This is primarily due to the fact that radiant heat transfer of furnace gases produced by oxyfuel combustion is significantly more efficient than those of air-fuel. Due to the absence of nitrogen in the combustion mixture, which does not need to be heated, the volume of exhaust gas is also substantially reduced, thus lowering total heat loss through the exhaust gas. Thanks to improved thermal efficiency, the heating rate and productivity are increased and less fuel is required to heat the product to a given temperature, while at the same time economizing on fuel and reducing CO_2 emissions.

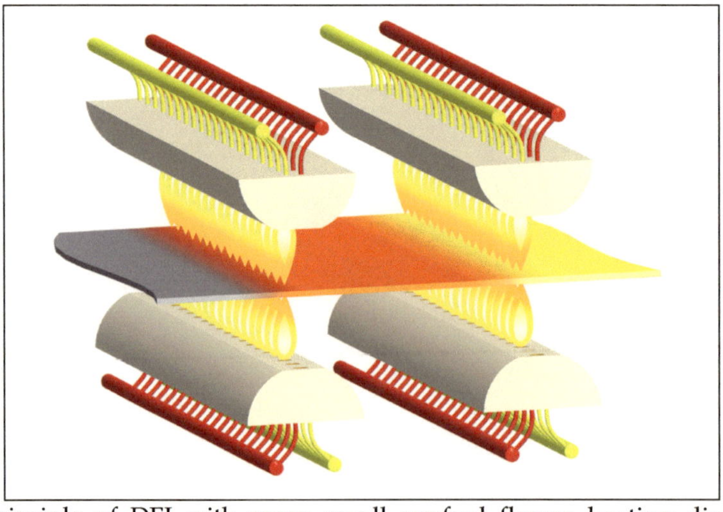

Fig. 5. The principle of DFI with many small oxyfuel flames heating directly onto the material.

The DFI unit has a thermal efficiency of around 80%. This reduces the specific fuel consumption while delivering a powerful 30% capacity increase in an existing strip processing line. In galvanizing, zinc adhesion and surface appearance are also improved due to DFI's effective pre-cleaning properties, leaving both strip and furnace rolls cleaner than before.
It is important to note that applying a DFI Oxyfuel system for preheating a strip does not create an oxidation problem; for example, experience with preheating up to 300°C shows no problems. In metal coating lines, the thin oxide layer formed is reduced in the subsequent reduction zone. It is also possible to influence the oxidation level to a certain extent by adjusting the stoichiometry of the flames.

3. At the integrated iron and steel processes

To produce iron in solid or liquid form from iron ore a reductant is needed. The two most suitable reductants, from a technical and economic perspective, are carbon and hydrogen. The use of pure hydrogen is today frequently not realistic; there are a few exceptions but these are linked to unusual localised conditions. In practice, the reductants used in today's iron-making are coal and natural gas; the coal being in the form of coke or pulverized coal. Blast furnaces inevitably require coke to a certain extent. It should be noted that the lowest operational limit of coke in a shaft furnace has been estimated at 150-200 kg/t. This is determined by the requirements for the carburization of iron, direct reduction with carbon and, in particular, shaft permeability and burden support.
To improve throughput and decrease of CO_2 emissions, so-called Full Oxygen Blast Furnace processes are frequently discussed as a possible alternative. The idea of the Full Oxygen Blast Furnace (FOBF) is not new; some researchers discussed it back in the 1930s and 1950s. The modern ideas were presented in the 1980s. They are based on two main principles: using pure oxygen as blast instead of air to create oxyfuel combustion of the injected pulverized coal, and acquiring a pre-reduction degree of iron (for example, 90%). However, FOBF processes are hampered by the so-called "hot bottom and cold top problem". Since Fink presented a proposal for an FOBF process in a patent in 1978, the idea of using recirculated top gas for compensating the decreasing amount of shaft gas and adjusting the very high flame temperature has been the basis for most other proposals.
Lately, FOBF processes - which have not yet been taken into full-scale operation - have experienced a renaissance, seen by many as the best way to decrease CO_2 emissions. The potential benefit of FOBF processes lies in the possibility of achieving a top gas with low nitrogen content (with a calorific value of 2 kWh/Nm^3, more than twice that of conventional blast furnaces) from which the CO_2 can then be removed reasonably effectively. The top gas is then recirculated back into the furnace as part of the fuel input. The captured CO_2 can be disposed of so that it is not discharged into the atmosphere or, for example, used in oil and natural gas fields. Using FOBF processes is a possible solution, but it is not yet a proven alternative.
Large benefits can be achieved from improved utilization of gases from other facilities on site, like coke oven gas from the coke-making and BOF gas from the steel-making converter (Basic Oxygen Furnaces - BOF). As the energy content of those gases are rather low, from half that of natural gas and downwards, combustion with oxygen is very beneficial or even

necessary. This could take place when injecting into the blast furnace or when using them as fuel in different types of heating operations. This also applies to use of blast furnace top gas.

Fig. 6. Blast furnace hot stoves, the large heat exchangers for heating the air-blast to over 1000°C prior to injection into the blast furnace tuyères.

An area where increased use of blast furnace top gas could be very beneficial is at the hot stoves. Use of flameless oxyfuel in blast furnace hot stoves is now under way. An evaluation of applying the technology in a large modern and efficient blast furnace, which produces a low calorific top gas, includes the following key observations:

- 25% of flue-gas can be recycled and this leads to a modified flue-gas composition containing 60% CO_2, accordingly increasingly suitable for Carbon Capture and Sequestration.
- The heat transfer by radiation is increased by 15% relative to conventional practices and this will manifest as improved stove efficiency.
- 60% of the fuel gas enrichment can be eliminated.
- The total energy use at the hot stove is reduced by 5%.
- The temperature of the blast increases by about 15°C.

Use of flameless oxyfuel in blast furnace hot stoves could, thus, replace combustion of coke oven gas or natural gas in this process with blast furnace top gas. This would typically cover the cost for the oxygen, or even provide a minor cost saving. The coke savings arising from an increase blast temperature will be substantial. In addition to the energy and environmental benefits, the stoves campaign life will be increased.

4. At the electric arc furnace

Today's electric arc furnace (EAF) for producing steel from scrap can be considered as a very sophisticated piece of equipment. During the 20th century, the development of electric steelmaking has been tremendous. In 1910, the electric furnaces, including both EAFs and induction furnaces, produced 0.2% of the world steel production, today this figure is nearly 35%. The two main factors explaining this evolution are the increased scrap availability and the development of ladle metallurgy, especially with the introduction of the ladle furnace (ASEA-SKF in 1965), which made it possible to increase both the production rate and the product quality. We should also bare in mind the favour of a much lower capital requirement as compared with the integrated steel mills.

When operating a modern EAF, the energy turnover is about 650 kWh/t, but only 60% of that energy is needed to heat and melt the scrap. A decrease of the energy turnover as such can be a goal, but many EAF mills consider a decrease of the electricity consumption as a more important way to cut costs and to enable an increased production rate. Decreased electricity consumption also offers additional advantages, such as lowered costs for electrodes and less disturbances on the grid. Moreover, electricity prices are at many places increasing. The electricity consumption can be lowered either by decreasing the total energy turnover or by replacing the electricity with energy in another form. It should be noted, that the electrical transmission losses are a direct function of the electricity consumption, representing >6% of the electricity supplied in an AC furnace and even more in a DC furnace.

The increasing use of oxygen has been very important for the development of EAF steelmaking. It begun with the (manual) oxygen lancing, in a first step used to replace the iron ore added during the refining period, but via oxyfuel burners and post-combustion it has developed into a number of more and more sophisticated applications. Today there are EAFs with a specific oxygen use above 50 Nm^3/t, more than the Basic Oxygen Furnace (BOF) in integrated steel mills.

The average ratio between the electricity savings and the oxygen use, should be about 3.5 $kWh/Nm^3\ O_2$. When introducing oxygen into an EAF, oxyfuel burners and oxygen lancing are employed in a first stage up to a total use of some 20 to 25 Nm^3/t usually with savings in electricity of about 5 $kWh/Nm^3\ O_2$ or more and with a corresponding increase of the production rate. When evaluating the overall reaction for oxygen lancing, ($C+\frac{1}{2}O_2=CO$), one should expect electricity savings to be maximum about 2 $kWh/Nm^3\ O_2$ even taking into account the higher contribution from dissolved carbon in steel scrap and adding energy corresponding to a possible post-combustion value of 8% in the bath-slag system. However, the much higher savings actually achieved, can be explained as follows. The overall reaction takes place in two steps: (1) the injected oxygen immediately combines with iron to form iron oxide, a strongly exothermic reaction, and (2) iron oxide in the slag is reduced by carbon, an endothermic reaction. The first reaction releases almost four times more heat per $Nm^3\ O_2$ than the overall reaction and this heat will be absorbed by surrounding scrap and significantly speed up the melt-down process.

Operating an EAF with under-pressure and especially with the slag door open during most of the operation leads to a heavy in-leakage of air. The oxygen part of this air could of course be of use inside the furnace, but the nitrogen (and argon) part is only to be considered as ballast. The energy demand for heating-up the ballast nitrogen, due to in-leakage of air, is 50-60 kWh/t. Even much higher figures, above 100 kWh/t, have been found at several EAF shops. The solution to this is to keep the slag door shut during the main part of the operation and run the furnace with a slight overpressure.

Since oxygen lancing was introduced, it has at most EAF shops been carried out by lancing through the slag door. Even if this way of lancing allows moving the injection point in all directions, the oxygen introduced will not be equally distributed throughout the bath. The trend of the EAF becoming more and more air-tight and the dynamic impact of a shorter meltdown time made it increasingly harder to use the conventional way of lancing oxygen and coal through the slag door. Nowadays we find combined equipment including all the functions: oxygen lancing, coal lancing, oxyfuel burners, and post-combustion. This equipment can be considered as combined lance-burners often with a coherent jet function enabling high-velocity injection, with a device for coal injection, where the burner also can be run overstoichiometrically to provide post-combustion or there is a separate nozzle for oxygen injection. To secure a good distribution of the heat supply throughout the furnace, including also the rear end of an Eccentric Bottom Tapping (EBT) type furnace, and the advantage of combining oxygen injection with oxyfuel burner operation, we end up with a minimum of four wall-mounted injection devices (assuming an AC furnace) - one at each cold spot between the electrodes and one in the EBT area.

The main factor limiting the energy supply from oxyfuel burners in an EAF is the heat transfer efficiency, which decreases with increased scrap temperature - we here have to compare with heat transfer from the electric arcs. However, as long as this heat transfer efficiency enables a decreased average cost for the production, it is of course beneficial to run the oxyfuel burners. Generally speaking, this normally means operation of the oxyfuel burners during about half of the time needed for the melting of each bucket of scrap charged, but the time is also a function of the production rate demand.

The CO/CO_2 ratio in equilibrium with liquid steel is high, even at low carbon contents. This result in a CO-rich gas leaving the bath-slag system in the furnace providing a potential for large energy recovery if this CO can be burnt with O_2 into CO_2 and the heat released be transferred to the metal. To illustrate the potential of post-combustion, we can say that in an EAF operation with a high coal injection, the energy released from the formation of CO is about 25 kWh/t, but if this entire CO can be transferred into CO_2 the total amount of energy released will be about 140 kWh/t. This should preferably be done with pure oxygen in order to minimize losses to the flue-gases.

Electricity savings from post-combustion are in the range 3-5 kWh/Nm^3 O_2, and can be obtained with rather simple means such as oxygen injection at fixed flow rates through existing oxyfuel burners during fixed periods of time, or by running the oxyfuel burners overstoichiometrically. For reaching high values, oxygen flow control through on-line flue-gas analysis and separate post-combustion lances can be used, making a heat recovery of 60-75% reasonable.

5. At vessel preheating

The use of oxyfuel to preheat vessels such as torpedoes, ladles and converters has been around for several decades. However, the number of installations is still surprisingly low given its potential. Using oxyfuel instead of air-fuel would reduce the fuel consumption drastically by approximately 50%, which would bring about a proportional decrease in CO_2 emissions. However, it would also have additional benefits such as a shorter heating time and hotter vessels. These would, for example, lead to fewer ladles in circulation and the possibility of reducing tapping temperatures. The latter directly saves energy in the furnace, but it could also decrease the tap-to-tap time of the furnace. The time saving would lead to additional energy savings as the specific (time dependent) heat losses from a furnace, would then be lowered.

If an oxyfuel ladle preheating system is installed adjacent to the EAF, preferably just a few metres away from the tapping position, very hot ladles can be used. Experience shows that such a measure would allow 20 minutes decreased ladle cycle and a 15°C lowered EAF tapping temperature, providing electricity savings at 5-6 kWh/t.

Let us look at a proven example of what this could lead to. The operating power with oxyfuel for a 60t ladle is approximately 1.2 MW. The average annual level is 0.8 MW, which at 7,500 h/y means 6 GWh/y. This is around half of what would be required with air-fuel; thus the annual saving is 6 GWh. Assuming the fuel is natural gas, the resulting decrease in CO_2 emissions would be 1,200 t/y, and this is only for one preheating station; normally there are multiple at each site.

Fig. 7. Ladle preheating using flameless oxyfuel at Ovako's Hofors Works, Sweden.

Conventional oxyfuel delivers a simple, compact and low weight installation as compared to an air-fuel system with a recuperator or regenerative solution. However, in preheating of vessels flameless oxyfuel brings additional strong advantages. Flameless oxyfuel is seen as the best available technology for heating and not only allows for ultra low NO_X emissions, but brings extended refractory life through more uniform temperature distribution. The first installation took place in 2003. Today more than 15 installations of flameless oxyfuel are in operation, two recent cases are found at Outokumpu at Tornio, Finland and SKF at Katrineholm, Sweden.

In 2008 flameless oxyfuel preheating was installed at Outokumpu's 90 tonnes ferrochrome converter. The 2.5 MW flameless oxyfuel system is used for drying and preheating of the converter, and provides the Tornio Works with greater energy efficiency, lower fuel consumption, and reduced emissions CO_2 and NO_X.

At SKF a similar type of flameless oxyfuel technology was installed last year, but for preheating ladles. And the size is here completely different; the ladles are for just 1 tonne of steel. Six ladle preheating stands were equipped with OXYGON® flameless oxyfuel preheating systems. This installation shows that a new energy saving and environmentally friendly technology also can be viable in a smaller scale production.

6. At reheating

Prompted by rapidly rising fuel prices in the 1970s, the steel industry began to consider methods to reduce fuel consumption in reheating and annealing. This laid the foundation for the use of oxyfuel solutions in rolling mills and forge shops. In the mid 1980s, some of

these furnaces got equipped with oxygen-enrichment systems, which increased the oxygen content of the combustion air to 23-24%. The results were encouraging: fuel consumption was reduced and the output, in terms of tons per hour, increased.
Oxyfuel solutions deliver a unique combination of advantages in reheat and annealing. Thanks to improved thermal efficiency (about 80% compared with 40-60% for air-fuel), the heating rate and productivity are increased and less fuel is required to heat the product to the desired temperature, at the same time saving on CO_2 and NO_X emissions. In summary the results include:

- Throughput capacity increase of up to 50%
- Fuel savings of up to 50%
- Reduction of CO_2 emissions by up to 50%
- Reduction of NO_X emissions
- Reduction of scaling losses

In 1990, Linde converted the first steel reheating furnace in the world to operate with 100% oxygen at Timken in the USA Since then, Linde has been pioneering the use of oxyfuel for this application. Today there are 120 reheat furnaces and annealing lines using Linde's oxyfuel solutions. The best air-fuel solutions need at least 1.3 GJ for heating a tonne of steel to the right temperature for rolling or forging. When using the REBOX oxyfuel solutions the comparable figure is below 1 GJ, a saving of 25%. For continuous heating operations it is also possible to economically operate the furnace at a higher temperature at the entry side of the furnace. This will even further increase the possible throughput in any furnace unit. Oxyfuel combustion allows all installation pipes and flow trains to be compact without any need for recuperative or regenerative heat recovery solutions. Combustion air-blowers and related low frequency noise problems are avoided.

		Air-fuel	Air-fuel with recuperator	REBOX® oxyfuel
Enthalpy in steel	kWh/t	200	200	200
Transmission losses	kWh/t	10	10	10
Flue-gas enthalpy	kWh/t	290	155*	50
Flue-gas temperature	°C	1,200	850	1,200
Air preheating	°C	20	450	20
Thermal efficiency	%	42	60	80
Energy need	**kWh/t**	**500**	**365**	**260**
Energy need	**GJ/t**	**1.8**	**1.33**	0.94
Oxygen production	kWh/t			25

*after recuperation

Table 1. With oxyfuel it is possible to achieve an 80% thermal efficiency, as compared with 60% in the best air-fuel cases. Even if also adding the energy needed to produce the required oxygen, we would reach 285 kWh/tonne, thus still close to 1 GJ, a saving of 20%.

During the last years flameless oxyfuel have been employed, for example in Brazil, China, France, Sweden, and the USA. Here follows some examples from those installations.

Soaking pit furnaces at Ascométal

There are flameless oxyfuel installations at two sites belonging to the bearing steel producer Ascométal in France, which is part of the Severstal Group. At Fos-sur-Mer, a turnkey delivery in 2005-2007 converted nine soaking pit furnaces into all flameless oxyfuel. The delivery included a combustion system with flameless burners, furnace upgrade, new flue-gas system, flow train, and a control system. The furnace sizes are 80 to 120 tonne heating capacity each. The results include 50% more heating capacity, 40% fuel savings, NO_X emission reduced by 40%, and scale formation reduced with 3 tonne per 1,000 tonne heated. In a second and similar project in France in 2007-2008, four soaking pit furnaces at the Les Dunes plant were also converted into all flameless oxyfuel operation.

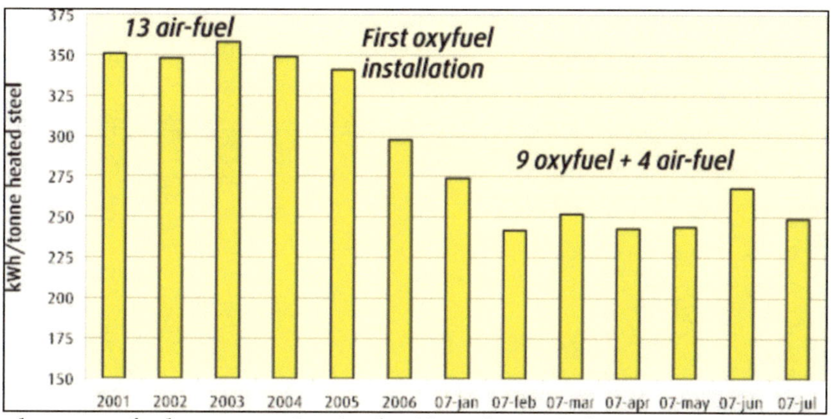

Fig. 8. Total average fuel consumption in the 13 soaking pit furnaces at Ascométal, Fos-sur-Mer. 2001-2004 was all air-fuel combustion. The first conversion into oxyfuel took place in 2005. In 2007 nine out of 13 furnaces were operated with all oxyfuel. The average fuel consumption per tonne for all furnaces was reduced by 100 kWh or 10 Nm^3 of natural gas.

15 installations at Outokumpu

At Outokumpu's sites in Sweden there are a total of 15 installations. In 2003, a walking beam furnace in Degerfors was rebuilt and refurbished in a Linde turnkey project with performance guarantees. It entailed replacing the air-fuel system, including recuperator, with flameless oxyfuel, and installation of essential control systems. The resultant 40-50% increase in heating capacity provided increased loading of the rolling mill, reduction of over 25% in fuel consumption and NO_X emissions below 70 mg/MJ.

At the Nyby plant, there are two catenary furnaces, originally installed in 1955 and 1960 respectively. The catenary furnace on the first annealing-pickling line, for hot or cold rolled strips, was converted to all oxyfuel operation in 2003. Requirements for increased production combined with stricter requirements for low NO_X emissions led to this decision. The furnace, 18 m long, was equipped with flameless oxyfuel burners. The total power input, 16 MW, was not altered when converting from air-fuel to oxyfuel, but with oxyfuel the heat transfer efficiency increased from 46% to 76%. The replacement of the air-fuel system, combustion blowers and recuperators resulted in a 50% increase in heating capacity

without any increase in the length of the furnace, a 40% reduction in specific fuel consumption and NO_X levels below the guaranteed level of 70 mg/MJ.
At Avesta we find the world's largest oxyfuel fired furnace, 40 MW. The old 24 m catenary furnace had a 75 tph capacity, but the requirement was to double this whilst at same time meeting strict requirements for emissions. The refurbishment included a 10 m extension, yet production capacity was increased to 150 tph. The conversion involved the removal of air-fuel burners and recuperators and the installation of all oxyfuel. The oxyfuel technology used involved staged combustion. The conversion reduced fuel consumption by 40%, and NO_X levels are below 65 mg/MJ. This furnace is an example of another route to flameless; having been converted from conventional oxyfuel to flameless oxyfuel last year and resulting in an additional 50% reduction of the NO_X levels.

Fig. 9. A heated slab is discharged from the walking beam furnace at Outokumpu's Degerfors Works. Here flameless oxyfuel has increased the heating capacity by 40-50%.

50% fuel savings at ArcelorMittal
There have been several successful installations in rotary hearth furnaces. One is found at ArcelorMittal Shelby in Ohio, USA. In 2007, Linde delivered a turnkey conversion of a 15-metre diameter rotary hearth furnace at this seamless tube producer. It included combustion system with flameless burners, furnace upgrade, new flue-gas system, flow train, and a control system. The former air-fuel fired furnace was converted in two steps, first using oxygen-enrichment for a period of time and then implementation of the flameless oxyfuel operation. Excellent results have been achieved, meeting all performance guarantees. These included >25% more throughput, 50% fuel savings compared with oxygen-enrichment (60% below the prior air-fuel performance), CO_2 emissions dropped accordingly, NO_X emission <70 mg/MJ corresponding to 92% less on an annual basis, and 50% reduced scale formation.

In January 2010, ArcelorMittal's received the Association for Iron & Steel Technology's Energy Achievement Award for its efforts to reduce fuel consumption and emissions using the REBOX oxyfuel technical solution at its Shelby mill.

Fig. 10. Outside view of the rotary hearth furnace at ArcelorMittal Shelby after the conversion into all flameless oxyfuel operation. Please note that all bulky equipment and piping relating to the previously used air-fuel system have been removed as it no longer is of any use.

SSAB Walking Beam Furnace with REBOX HLL

At SSAB in Sweden REBOX HLL is used. The slabs are reheated in walking beam furnaces with a capacity of 300 tph per furnace, from ambient temperature to 1,230°C. The air-fuel combustion system uses a recuperation system to preheat air to 400°C. The fuel is oil, and prior to the HLL installation the consumption was 440 kWh/tonne, or 1.58 GJ/tonne.

REBOX HLL creates a type of flameless oxyfuel without replacing the existing air-fuel burners. By reducing the air flow and substituting high velocity oxygen injection into the combustion, great benefits can be achieved. 75% of the oxygen needed for the combustion is supplied with this technique. The flue-gas volume is less than 45% that of air-fuel. The system start-up took just one day. The installation in only one zone has increased the heating capacity from 300 to 320 tph.

The installation of HLL is rather easy because it does not include any replacement of burners or installation of additional burners, which minimizes the installation down time. The air-fuel system can at any time be brought back into operation as it was before. This eliminates any potential risk relating to the implemention, and it enables operation to be flexible and optimized in response to fluctuating fuel cost and production requirements.

Some important results from this installation are:

- No negative impact on the surface quality.
- A positive impact on the temperature uniformity of the slabs.
- The ideal heating curve suggested by the control system can be achieved more easily.
- Less smoke emanating from the furnace, greatly improving the plant environment.

Based on the results of current installation in one zone, SSAB has estimated that a full implementation would provide the following:

- A reduction of NO_X emission by 45%.
- Fuel consumption can be decreased by 25%, leading to the same reductions in SO_2 and CO_2 emissions.
- Production throughput can be increased by 15-20%.

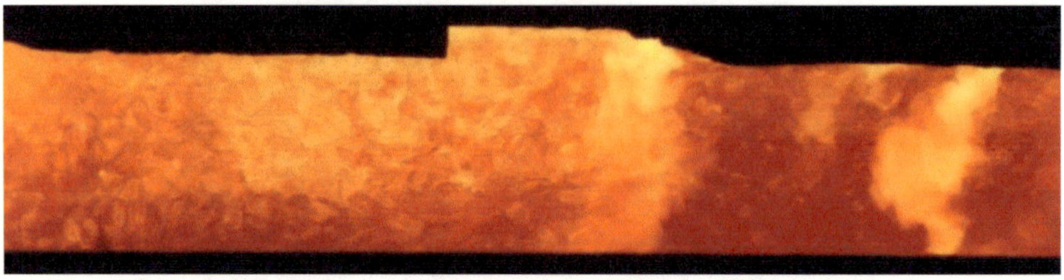

Fig. 11. "Semi-flameless" oxyfuel combustion in a 300 tph walking beam furnace at SSAB, Sweden.

Stainless wire annealing in China

At Dongbei Special Steel Group in China, a new state-of-the-art annealing furnace for stainless steel wire has been taken into operation in 2010. It applies a combined technology called REBOX DST (Direct Solution Treatment), the benefits compared with a conventional solution are extremely huge, for example the treatment time is drastically reduced. The flameless combustion here uses a low calorific fuel with an energy content of 1.75 kWh/Nm^3 (6.3 MJ/Nm^3).

7. At strip processing

Flameless oxyfuel can be used for heating at strip processing, but the real difference here is made by applying DFI Oxyfuel, a fascinating, compact, high-heat transfer technology, which provides enhanced operation in strip processing lines such as galvanizing. DFI Oxyfuel has been used to boost capacity of strip annealing and hot dip metal coating lines by 30% or more, while reducing the specific fuel consumption. Systems are in operation at Outokumpu's Nyby Works in Sweden and ThyssenKrupp's works at Finnentrop and Bruckhausen in Germany. In mid 2010 a unit was installed in a continuous annealing line at POSCO in Pohang, South Korea.

Since the beginning of the 1990s, Linde has pioneered the use of 100% oxyfuel applications in reheat furnaces in close cooperation with customers such as Outokumpu. At Outokumpu's Nyby site in Sweden, the company wanted to increase the production capacity of a stainless strip annealing line, but the furnace already contained an oxyfuel combustion system and had extremely limited physical space available. In 2002, the first compact DFI Oxyfuel unit was installed, making it possible to increase the production by 50% (from 23 to 35 tph) without extending the furnace length. This DFI Oxyfuel installation consisted of a 2-metre long DFI unit at the entry side with four burner row units including a total of 4 MW installed power distributed on 120 oxyfuel flames.

In 2007, the REBOX DFI system was installed at ThyssenKrupp Steel's (TKS) galvanizing and aluminizing line in Bruckhausen, Germany. Earlier, Linde had installed a DFI unit at the TKS galvanizing line at Finnentrop, and increased production from 82 to 105 tph, or over 30%. The results at the Bruckhausen installation matched those in Finnentrop: increasing capacity from 70 to 90 tph. Oxyfuel not only effectively heats - contributing to a reduction of fuel consumption - but also cleans, thus eliminating the need for the pre-cleaning section. In addition, the process made it possible for ThyssenKrupp to pre-oxidize steel strips in a precise and controlled manner. Prior to the DFI installation, the Finnentrop plant had a 25 m long pre-cleaning section with electrolytic cleaning and brushes.

At Finnentrop, to minimize line downtime, the design resulted in a 3-metre long DFI unit equipped with four burner row units, with a total of 120 oxyfuel flames and 5 MW installed power, with an option of two more row sets for an additional 2.5 MW. Three metres of the existing recuperative entry section was removed to fit the DFI Oxyfuel unit. The number of burner row units and burners employed depend on set preheating temperatures and the actual strip width and tonnage. At 105 tph, DFI Oxyfuel results in an immediate steel strip surface temperature increase of more than 200°C.

With the DFI unit the capacity of the Finnentrop line increased from 82 to 109 tph. The DFI Oxyfuel unit also manages to burn off residue, particles, grease and oil from the strip rolling process, providing a cleaner strip than the long electrolytic and brush strip pre-cleaning section used to do. At a production level of 36,000 tonnes per month at Finnentrop, results include an over 5% reduction in natural gas consumption, almost 20% less NO_X emissions, and a reduction of 1200 tonnes per year in CO_2 emissions.

Fig. 12. REBOX DFI installation in a galvanizing line at ThyssenKrupp Steel at Finnentrop, Germany. The 3-metre long DFI unit was fitted into the previous (non-fired) dark-zone.

The oxidation is lower than normal at a specific strip temperature since the dwell time is very limited; applying DFI Oxyfuel for preheating a strip up to 300°C does not create oxidation problems. In metal coating lines, the thin oxide layer formed is reduced in the subsequent reduction zone. It is also possible to influence the oxidation by adjusting the stoichiometry of the flames, for example by changing the lambda value from 1.0 to 0.9.

The oxide layer thicknesses have been measured to be in the range of 50-100 nanometres, even at high strip temperatures. A well performing reduction zone should be able to reduce the scaling further. For high strength steel, a small formed oxide layer, for instance, 200 nm, may be beneficial, since after reduction in the Radiant Tube Furnace section, pure iron will form on the surface for improve zinc adhesion.

Cleaning tests show that the carbon and iron fines contaminations can be drastically reduced by use of DFI. With the DFI Oxyfuel technology the cleaning section can be shortened to a spray cleaning section, one brush machine and a final rinsing section. The final cleaning operation is transferred to the DFI Oxyfuel inside the thermal section. The elimination of one brush machine and the electrolytic cleaning section brings considerable cost savings in maintenance and operation due to energy savings and less wear parts. Furthermore, DFI gives potential to reduce investment and operating costs in the furnace section since the furnace length can be reduced; the preheating and one heating zone can be saved.
This year, 2010, REBOX DFI is for the first time employed in a continuous annealing line for carbon steel, at POSCO's large integrated steel mill at Pohang, South Korea. The DFI unit provides a guaranteed level of preheating which will be capable of achieving approximately 15% higher capacity in the annealing furnace. The natural gas fired DFI unit consists of four oxyfuel burner row units with a combined capacity of close to 6 MW.

8. Opportunities for decreasing CO_2 emissions

There is a strong political will to decrease CO_2 emission. The steel industry only accounts for some 3% of worldwide CO_2 emissions, which totals roughly 30 billion tonnes per annum relating to the human activity of burning of fossil fuels, but seems to be strongly affected by this. To radically change existing processes and production routes to decrease the CO_2 emissions would be extremely expensive, even if it were possible.
However, there exist today a number of proven solutions and technologies which, if fully implemented, could substantially decrease CO_2 emissions without seriously altering current methods of operation and are therefore short-term viable solutions. If these solutions are fully implemented, the combined impact on CO_2 emissions from the steel industry worldwide is estimated to be a reduction of 100 million tonnes of CO_2 per annum within a relatively short time span. Among these solutions, the most viable is oxyfuel combustion.

Fig. 13. A look through the furnace door of the rotary hearth furnace at ArcelorMittal Shelby, USA; a flameless oxyfuel burner is firing straight towards the open door. Here the conversion from air-fuel to flameless oxyfuel led to a 60% reduction of the CO_2 emission.

CO_2 emissions from the steel industry have two main sources: reduction processes, and melting and heating processes. It is well known that reduction processes are the dominant source. The two main routes for steel production account for quite different impacts on CO_2 emissions: integrated steel mills, including all upstream processes, average approximately 2 tonnes of CO_2 per tonne of hot rolled plate; for mini-mills, the corresponding figure is 0.5-0.6 tonnes. However, the contribution from heating processes is not negligible; each piece of steel is on average heated twice on its journey through the production chain, and this is far from the only heating process. Accordingly, by increasing the energy efficiency in the heating processes, a large impact can be made on reducing the carbon footprint. An additional advantage is the low flue-gas volumes with high concentration of CO_2, which enable directing it to capturing and potentially sequestration.

Use of a fuel with a low calorific value is of interest in this context. It could, for example, be internally produced gas streams at a plant, like blast furnace top gas and BOF gas. In many places, at least some of the latter gases are not used but put to flaring. What is frequently hampering their greater use is the flame temperature required in heating applications. However, using oxyfuel instead of air-fuel would in many cases make it possible to even run solely a low calorific gas as fuel. Where these gases are being flared today, the resultant impact on the site's CO_2 emissions of using them in this way would be very positive and would replace other energy sources. A practical example of an increased use of a low grade fuel can be found in blast furnace hot stoves, where due to the oxygen-enrichment it leads to improved fuel economy and reduced CO_2 emissions.

As the examples and solutions discussed in this chapter all use oxygen, it is appropriate to comment on the CO_2 emissions relating to oxygen production. The production of 1 Nm^3 of gaseous oxygen requires approximately 0.5 kWh of electricity. If this electricity is produced by hydro or nuclear power plants, it "carries" no CO_2. However, if produced using fossil fuel it would correspond to 0.5 kg CO_2 per Nm^3 of oxygen. Thus, in the worst case scenario, oxyfuel combustion contributes (from oxygen production) 0.1 kg CO_2 per kWh. Turning that worst case scenario into practice, it is known that oxyfuel combustion (compared with air-fuel) would reduce the fuel consumption by an average of 40%, and the combined effect on CO_2 emissions would then be a reduction of approximately 35%.

9. Conclusions

The traditional use of oxyfuel in steel-making is in the electric arc furnace. Today sophisticated wall-mounted equipment is used combining the functions of oxygen and coal lancing, oxyfuel burner, and post-combustion. The level of oxygen use could reach above 50 Nm^3/t, more than in the steel-making converter in integrated steel mills.

Mainly due to the strive to reduce CO_2 emissions the Full Oxygen Blast Furnace concept is now being tested. Here oxygen is completely replacing the air-blast. However, in a short-term perspective it seems advantageous to instead focus on the hot stoves, where low calorific fuel can be used to an increased extent, a typical benefit from oxyfuel.

Oxyfuel provides an overall thermal efficiency in the heating of 80%, air-fuel reaches 40-60%. With flameless oxyfuel, compared to air-fuel, the energy savings in a reheating furnace are at least 25%, but many times 50% or even more. It is possible to operate a reheat furnace with fuel consumption below 1 GJ per tonne. The corresponding reduction in CO_2 emissions is also 25-50%. Savings in terms of NO_X emissions are substantial. Flameless oxyfuel combustion has major advantages over conventional oxyfuel and, even more, over any kind of air-fuel combustion. The improved temperature uniformity is a very important benefit, which also reduces the fuel consumption further.

With oxyfuel it is possible to increase the throughput rate by up to 50%. This can be used for increased production, less number of furnaces in operation, increased flexibility, etc. It is also of interest when ramping up production; two furnaces can cover the previous production of 2.5-3 furnaces, meaning possibility to post start-up of the third furnace and, additionally, resulting in decreased fuel consumption. Increased capacity can also be used to prolong soaking times. Thanks to the reduced time at elevated temperatures, oxyfuel leads to reduced scale losses, at many installations as high as 50%.

Using DFI Oxyfuel, where the flames heat directly onto the moving material, a very compact solution has been established. Installations show the production throughput can be increased by 30%, but it also provides other important benefits. This technology is particularly suitable for strip processing.

The experiences from converting furnaces into all oxyfuel operation show energy savings ranging from 20% to 70%, excluding savings in energy needed for bringing the fuel to the site. The use flameless oxyfuel in ladle and converter preheating is extremely advantageous. Now we also see that this innovative technology can be used at blast furnace hot stoves to improve energy and production efficiencies and reduce environmental impact.

There exist today a number of solutions and technologies which could substantially decrease CO_2 emissions without seriously altering current methods of operation and are therefore short-term viable solutions. Additionally, they would lead to improved fuel economics and reduced processing times. In heating and melting, oxyfuel combustion offers clear advantages over state-of-the-art air-fuel combustion, for example regenerative technology, in terms of energy use, maintenance costs and utilization of existing production facilities. If all the reheating and annealing furnaces would employ oxyfuel combustion, the CO_2 emissions from the world's steel industry would be reduced by 100 million tonnes per annum. Additionally, a small off-gas volume and a high concentration of CO_2 make it increasingly suitable for Carbon Capture and Sequestration.

Using oxyfuel instead of air-fuel combustion for all kinds of melting and heating operations opens up tremendous opportunities, as it leads to fuel savings, reduces the time required for the process and reduces emissions. Numerous results from installations have proven this.

Low-energy buildings – scientific trends and developments

Dr. Patrik Rohdin[1], Dr. Wiktoria Glad[2] and Dr. Jenny Palm[2]
[1]*Energy systems, Linköping University*
[2]*Tema T, Linköping University*
Sweden

1. Introduction

Over the past twenty years primary energy demand in the world has increased drastically, while during the same time demand for electrical energy has increased even more. This, in combination with the impact of global warming, is forcing policy-makers to formulate goals to meet this threat. The EU Commission has recently stated that one of its highest priority tasks is to address global warming, with special focus on reducing greenhouse gases. The EU Commission states in the directive for energy efficiency in the built environment that the building sector must decrease its use of energy to reduce CO_2 emissions. In addition a goal for energy efficiency within the Union states that a 20% increase in energy efficiency shall be met by 2020. The Swedish parliament has also set a national goal for space heating, which states that by 2020 the use per floor area should be reduced by 20% and by 2050 this figure should be 50% compared to use during 1995.

To be able to meet these goals, many different activities must strive towards the same goal. One major part is the building and service sector, which accounts for about 35% of total Swedish national energy use. A large part of that use is concentrated in cities, which underlines the importance of working with such areas. The connection between CO_2 emissions and the use of energy is also an important motive for promoting a more efficient use of energy and reducing the total energy demand. This means that there is a need to choose the correct primary energy and energy conservation measures as well as to reduce the total electrical usage in the built environment.

Furthermore, the consequences of global warming are introducing changing conditions to be met by future buildings with increasing temperatures, and for Sweden increasing precipitation as well. In IPCC (2007) the temperature increase is predicted to be 1-2°C with an increase in precipitation by 20% for the 2020-2029 scenarios relative to 1980-1999. For the long-term scenario until 2090-2099 the predictions are of the order of 4-5°C. Effects like this should be included in the analysis of future energy systems and design criteria, since it will reduce heat demand and increase the risk of overheating in buildings. Poor indoor environmental conditions in buildings is an important factor which costs large amounts of money in healthcare and administration, while a well-functioning indoor environment plays an important part in a convenient and modern life. It is also important to include the environmental impact from building materials.

One key component in achieving a more sustainable building sector is to introduce different forms of energy-efficient or renewable buildings. In this chapter a review of the literature published within the Web of Science databases on low-energy houses, passive houses, zero-energy houses and passive solar houses is presented. The aim is to analyze trends in the scientific literature concerning sustainable buildings and to discuss which issues have been in focus and which have been neglected in earlier studies. This will create a basis for discussing knowledge gaps and future research needs. Our scope is to focus on the development of research on dwellings.

2. Field overview

The field of low-energy buildings is broad and complex. The first article included in this review is from 1978. A total of 83 relevant hits were found within Web of Science for the seven search words (1) low-energy buildings; (2) low-energy architecture; (3) low-energy house; (4) passive house; (5) passive solar building; (6) passive solar house; and (7) zero energy house. The number of unique hits for each search word is seen in Figure 4.

The trend of increasing interest in low-energy buildings can also be seen in the increase in production of scientific papers within the scope of this review, see Figure 1. During the last five years the number of publications has moved up from about one or two per year to between eight and ten. This should also be seen in the light of an increase in the general production of papers, but at the same time it shows that there is strong focus on low-energy solutions for the built environment in the scientific community too.

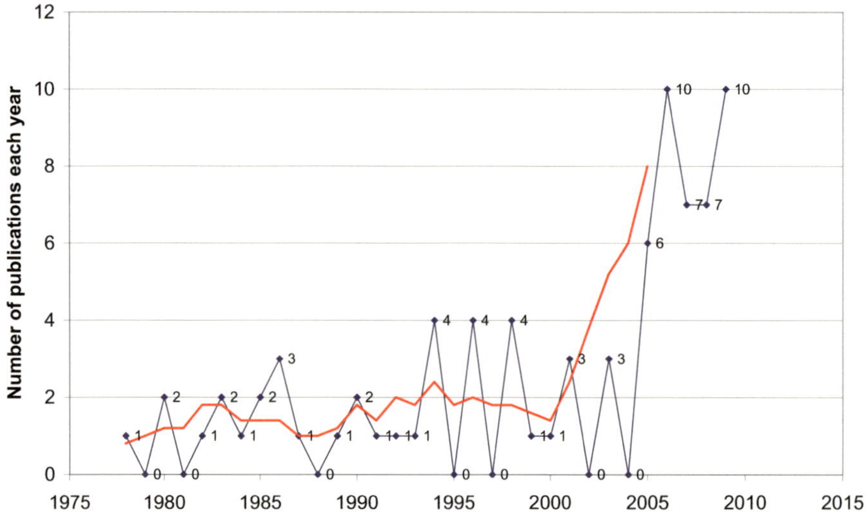

Fig. 1. Overview of the number of publications each year from the first article reported in a journal in Web of Science in 1978 to today (2010). The red line indicates the five-year moving average.

We have used the Web of Science database when we searched for relevant articles. This database is dominated by journals with a technical focus, which may partly explain that when examining the structure and focus of the reviewed articles in terms of the main method used it turned out to be a highly technical field. But we also noticed that within these journals examples of broader articles including policy issues, interdisciplinary studies and economic studies have become increasingly common in the last few years. But the main part of the field still remains technical in nature, with focus on building energy simulation (BES), component studies of thermal walls and solar applications and measurements, see Figure 3. There is also a strong tendency for the field to employ case studies and experimental setups either in laboratory form or as in real constructions, see Figure 2.

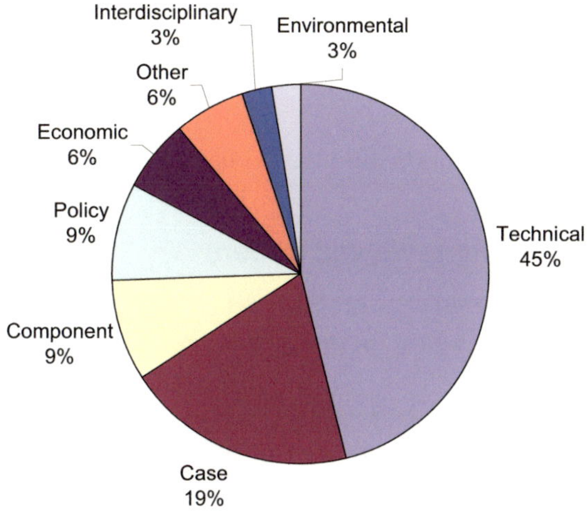

Fig. 2. A distinction between different types of article within the review. The categorization is not unambiguous since several articles may be relevant for more than one of the suggested groups.

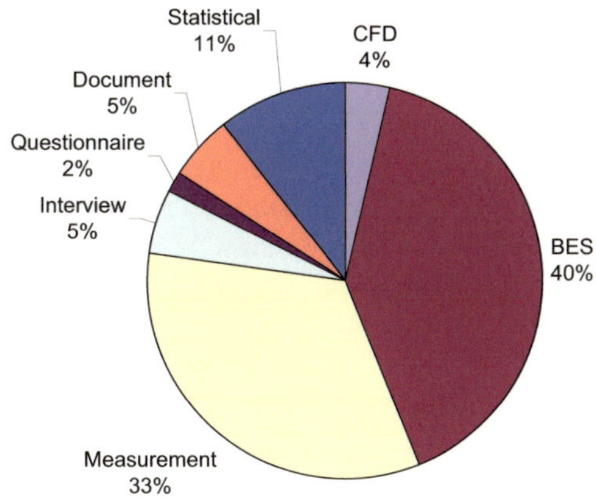

Fig. 3. The relative difference in number of publications using different methods. This distinction is however not unambiguous since several papers can be argued to have more than one of the above suggested methods.

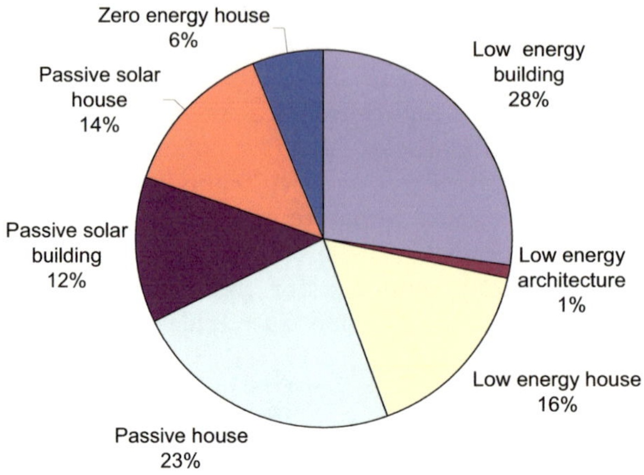

Fig. 4. The relative magnitude of hits within the review on different search words.

3. Main methods cited in the reviewed papers

This section will present an overview of the method characterisation used in the review and introduce the concepts of the different methods. The main methods identified in the articles are: (1) Computational Fluid Dynamics; (2) Building Energy Simulation; (3) Measurements; (4) Interviews; (5) Questionnaires; (6) Statistical; or (7) Environmental or Life Cycle-Focused Studies.

3.1 Computational Fluid Dynamics (CFD)

Computational Fluid Dynamics (CFD) has been extensively used as a scientific tool in many application and research situations since the 1950s. The use is widespread in many fields, such as aerodynamics, hydraulics, combustion engineering, meteorology, electronic cooling and biomedical engineering, and in predicting the external and internal environment of buildings. In Versteeg and Malalasekera (1995) the authors give a rather broad definition of CFD: "Computational fluid dynamics (CFD) is the analysis of systems involving fluid flow, heat transfer or associated phenomena such as chemical reactions by means of computer based simulation."

The use of CFD to simulate ventilation and air movements in rooms is becoming more and more common. One of the earliest publications where CFD was used to simulate air flow in rooms was made by Nielsen in 1974. Due to the increase in computer resources, the use of CFD as a scientific tool has increased and continues to increase as it is possible to solve more complex and challenging problems. As the cost and time needed to perform real experiments in many cases are high, CFD has become more and more extensively used. This method is of course of special interest for cases where it is not possible to obtain measurements, such as situations where the object has not yet been built. However, to ensure the validity and reliability of CFD models, measurements are still very much needed. An often-used approach is to compare results from numerical simulations with measurements; if the results coincide, a numerical approach in predicting similar situations may be used.

3.2 Building Energy Simulation (BES)

Building Energy Simulation (BES) is a frequently used tool to predict energy use in buildings within the academic sphere as well as in the design process in the construction industry. Similar to other types of simulations, BES is a numerical experiment using a mathematical model. The aim is to predict or forecast a future or an otherwise presently unknown situation. For energy simulation programs, issues such as predicting energy use, either in a future building not yet built, or after a change in a system has been made in a present building, are of interest. In Bergsten (2001) a comparison of different energy simulation software is presented, and a classification of the software is made depending on whether it is a general simulation program or has multi-zone capabilities and if it is static or dynamic. The software compared, considered the most important energy simulation software used in Sweden, Norway and Denmark, were Bsim 2000, BV2, EiB, IDA ICE, Energikiosken, Enorm 2000, Huset, OPERA, Villaenergi, VIP+ and Värmeenergi (Bergsten, 2001). In Crawley et al. (2005) a more extensive review of the performance and capabilities of building energy simulation programs is presented. The review includes BLAST, BSim, DeST, DOE-2. IE, ECOTECT, Energy-10, Energy Express, Ener-Win, EnergyPlus, eQUEST, ESP-r, IDA ICE, IES<VE>, HAP, HEED, PowerDomus, SUNREL, Tas, TRACE and TRNSYS.

3.3 Measurements

Studies of indoor climate and energy efficiency often include measurements of temperature, moisture, air velocity, turbulence intensity, carbon dioxide, radon and other pollutants in addition to power and energy. When measuring spatial distributions there is a problem with creating a comprehensive view as it is time-consuming and is costly. Measuring the climate in a room with arbitrary accuracy is virtually impossible because it would require too many data points. It is also true that it will be time consuming and expensive to measure over long periods of time. Measurements as evaluation instruments are of course invaluable, but the very nature of the measurement in itself does not give any idea of the future, as it only says something about the past. At this stage different types of models are needed in order to make statements about the future. All measurements are also affected by different measurement errors. These vary greatly depending on the type of equipment used and the manner in which measurements are made.

3.4 Interviews

Interviewing is a common data collection method in social science qualitative research, among with observations and document analyses. The aim of qualitative inquiries is to explore the qualities of phenomena and provide data to gain deeper understanding (Lincoln and Guba, 1985). Using interviews to acquire data is usually preceded by a process of letting the problem at hand determine what type of inquiry is suitable and how the problem is best explored. A structured interview could in some cases generate similar data as a questionnaire, while a more open-ended, semi-structured interview requires more attentiveness and flexibility from the interviewer but can provide detailed descriptions and interpretations of phenomena in the world (Kvale and Brinkmann, 2009). While quantative data concern more or less of a studied entity, qualitative data concern similarities or dissimilarities. Analysis of interviews is descriptive, but the purpose is to reach beyond the description of the questions in the interview. The analysis means that, through reflection, the researcher abstracts from the description and searches for patterns and dysfunctional ties in relation to earlier studies or theories (Kvale, 1996).

3.5 Questionnaires

Questionnaires are an important part of survey research since this is the most common data collection method. Structured interviews might sometimes be similar to questionnaires with standardized answers although they may be closed or open-ended. When the objective of the research is specific, for example to determine which factors influence a certain phenomenon, a statistical analysis of data from questionnaires might be used. Recently, assumptions taken for granted in survey research have been questioned (Krosnick 1999). Response rates, pretesting of questionnaires and questionnaire design are among the contested aspects of this research. Research has shown, for example, that low response rates might show more accurate results than studies with high response rates (Visser et al., 1996).

3.6 Document studies

Analyses of written texts are important research tools in, for example, content analyses. Other types of analyses using texts as the primary source of data are discourse and argumentation analyses. The various types might be used differently in different fields of research. For example, linguists use discourse analysis to find syntax and choice of words in relation to contexts and focus is on language as a power tool, while social scientists might focus on discourse analysis as a tool in itself to reveal social practice (Bergström and Boréus, 2005). This is often referred to as critical discourse analysis (Fairclough, 2003). These texts might be minutes, records, protocols, letters, articles, books, newspapers, journals found in libraries, on the internet, in public or private archives and files, etc. Some texts are superior to other data collection methods in terms of reliability since they are real-time documentations of a phenomenon, compared to interviews which construct phenomena during the interview. Also, large amounts of written texts enable researchers to do structural analyses in relation to a research question (Lincoln and Guba, 1985)

3.7 Statistical

Statistics is a whole scientific field with several subfields, but in general, statistical methods include ways to collect, process and draw conclusions from statistical figures to enable, for example, sampling and analyse differences (Körner et al., 1984). Focus is on empirical data and statistical methods are used in both natural and social sciences. Descriptive statistics include mapping and providing information, while inferential statistics aim to present causal explanations and factors that influence a certain phenomenon. The basis for explorations is elements in a population (Moses, 1986). Populations and elements might be individuals but can also refer to material or physical entities. A census is a study using whole populations, but usually a sample of a population is selected in order to limit the research due to resource constraints. Social and economic sciences use questionnaires or observations to collect data, while the most common data collection method in natural sciences is different types of observations. Data can be analysed using computer programs such as SPSS.

3.8 Environmental or Life Cycle-Focused Studies

Different types of environmental or life cycle-focused studies form an eclectic category of research. A definition of life cycle assessment is "compilation and evaluation of the inputs, outputs and potential environmental impacts of a product system throughout its life cycle" (Guinée, 2002). Industrial ecology covers a similar field of exploration (Frosch and Gallopoulos, 1989). Some of these methods claim to have a "holistic" approach but in reality it is impossible to include all environmental aspects "from cradle to grave" and methods with this approach might be criticized because they promise too much. Also, a "system" approach is common among the different methods and may receive the same critique (ISO , 1997). The definition of system boundaries is a constant issue of debate. Another critical point in these analyses is the validity of data (Cooper and Fava, 2006). Common concepts included in energy-related research with these approaches are "embodied energy" (Odum, 2007) and exergy (Dinçer and Rosen, 2007). Social issues and phenomena that are hard to put in figures are generally missing in these studies.

4. Articles by method

In this section the articles from the review will be presented under a characterization of the method used in the paper. As previously stated, this is not an unambiguous characterization but should be seen as a way of grouping the studies reviewed within this chapter. Below a short presentation of the studies is made and a discussion of the main lessons learned will follow.

4.1 Computational Fluid Dynamics

In this section only two studies are represented. The first study was made for conditions in Japan. A constructed CFD code is made to study the transient effects of a thermal storage wall. The authors suggest that this technique may be viable for low-energy houses with hybrid systems (Onishi et al., 2001).

In Karlsson and Moshfegh (2006) the authors present an overview and results from a low-energy building built in Lindås, Sweden. The authors use CFD to study the air velocity and temperature pattern in one of the rooms in the building. In addition, building simulation is used, in the form of ESP-r, to study dynamic effects. Furthermore the paper investigates the importance of high-performance windows, and concludes that this is an important factor, both to decrease energy use and for the thermal environment. The paper also discusses the problem of overheating during summer.

4.2 Building Energy Simulation (BES)

Building energy simulation is a commonly used tool for predicting energy use and other parameters for buildings and to conduct parametric studies in different forms. A total of 18 papers are cited in this section. In Ohanessian and Charters (1978) a thermal simulation of a solar passive house with a Trombe-Michel wall is presented. The technique as well as measurements and simulations are presented, and an energy savings potential of about 40% is demonstrated for the passive solar house.

In Clarke et al. (1998) an integrated model of a low-energy building is presented. The case study is a city centre building in Glasgow, where an optimum mix of low, passive and active renewable energy technologies is sought. The main method used is the well-known simulation software ESP-r. The paper describes results from simulation and outlines a method by which "the replication potential of beneficial outcomes can be assessed".
A methodology for computational support during all phases of a design process with a focus on energy is presented in Shaviv (1996). The paper combines a procedural simulation approach with a knowledge-based heuristic approach in one integrated system. The overall aim is to provide architects and other actors in the design phase with a tool that can be used throughout the entire process (Shaviv, 1996). Holm (1996) also focuses on architects and the status of low-energy architecture in South Africa is presented. Chlela et al. (2009) discuss the design phase and a new methodology for design is proposed. Instead of parametric studies of design criteria using building energy simulation to optimize building envelope and HVAC, a Design of Experiments approach is suggested. The methodology is tested on three French cities with cold, moderate and hot climate. The models show "rather good results" for the annual heating demand and final energy use for the building as a whole. However, less accurate results were obtained for the cooling demand. The authors point out that further improvements may be made on the model.
In Lomas (1996) an application study of thermal simulation programs for passive solar house design from the U.K. is presented. This type of simulation program has been extensively used within the passive solar program in the U.K, and is also used in: (1) domestic and non-domestic building design studies; (2) in the assessment of innovative material and design techniques; (3) development of design guidelines; and (4) the design and interpretation of building monitoring studies. This makes it important that the programs must be reliable when it comes to: (1) the changes in energy demand of the building when making changes to the building such as changing window size; (2) the energy savings as a result of a design change, this to be able to predict pay-off times or other investment criteria; (3) the absolute energy demand of the house and the internal temperatures, this to be able to compare with energy targets and for example the risk of overheating, etc. The paper presents among other things (1) a methodology for structuring inter-program variability studies; (2) an overview of the U.K. application study project; and (3) a proposal of a Simulation Resolution (SR) concept. The authors argue that the SR could be taken as an estimate for the best (smallest) value that can be achieved for U.K. domestic buildings. The value provides a basis for estimating the significance of thermal predictions by this type of transient simulations. In the study ESP-r and HTB2 and SERIES were used.
Kalogirou et al. (2000) use an artificial neural network to predict energy use in a passive solar building. According to the authors, the model presented was able to predict energy use with acceptable accuracy. The model also proved to be much faster than dynamic simulation programs.
Solar heating is central in Badescu (2005) where active solar heating systems for passive houses are investigated using simulation. The suggested systems were tested on a passive house, and a suggested control scheme is outlined. In Badescu and Sicre (2003a) a description of the case is made, which is based on measurements and input from the Pirmasens passive house in Germany. Detailed input in the form of standardized data for a typical German family is used. Badescu and Sicre (2003b) reports on a model for predicting the thermal behaviour of this passive house. The topic of the paper is evaluation of

renewable energy in the context of passive houses. The renewable energy alternatives in focus in the paper are (1) passive solar heating with large windows facing the south; (2) active solar collectors for space heating and heating of domestic hot water; and (3) ground heat exchanger to preheat the supply air. The authors argue that the computational effort of transient simulation for this type of problem is valuable.

Feist et al. (2005) introduce and summarize the Passive House Standard and results from the EU project "Cost Efficient Passive Houses as European Standards" (CEPHEUS). The aim, according to the authors, is to "provide an acceptable and even improved indoor environment in terms of indoor air quality (IAQ) and thermal comfort at minimum energy demand and cost". This is achieved by improving the thermal performance of the envelope in such a way that the heating system can be simplified, thus keeping costs at a minimum. One important factor of the high-performance envelope is that the temperatures on the inside surface are close to the room air temperature and thus the radiation asymmetry is small. This enables high thermal comfort by the use of supply air heating instead of conventional radiator systems that usually compensate for both down draught and radiation asymmetry. If the thermal properties of the wall are low enough it is also possible to only use the IAQ-based supply air for heating, without exceeding 50°C which is a possible temperature to supply air without complications. The thermal transmittance for the wall is proposed to be <0.15 W/m^{2}K for Central Europe; for air tightness a value of <0.6 h^{-1} at 50 Pa; and for heat exchanger efficiency >75%. For a climate in central Europe a requirement of less than 15 kWh/m^{2}a is also set and a maximum power demand of 10 W/m^2. A more detailed description of values for different building components is found in Feist et al. (2005). The CEPHEUS project includes 221 housing units from five countries that comply with the passive house standards. The aim of the projects is, according to the authors, "to demonstrate the technical feasibility (in terms of achieving the target energy performance indices) at low extra cost (target: compensation of extra investment cost by cost savings in operation) for a variety of different buildings, constructions and designs implemented by architects and developers in several European countries.". Results in the study show no correlation between types of heating system and mean indoor temperature. Supply air heating was found suitable for passive houses. A comparison between the passive houses with other newly built conventional buildings show a reduction in useful energy by 56%, final energy 52% and primal energy by 56%. The thermal comfort is reported to be good to very good for these buildings built in central Europe.

In Persson et al. (2006) the authors investigate the influence of window size on the energy balance of low-energy buildings. The aim of the paper is to "investigate how decreasing the window size facing south and increasing window size facing north" would affect energy demand. A building energy simulation tool (DEROB-LTH) was used in the study. The authors conclude that the size of the energy-efficient windows does not have any major influence on heating demand during the cold season. However, the authors show that window size is important during summer, as it will affect the solar gains during this season. The main conclusion of the paper is that, according to the authors, it is possible to build windows in a more traditional way even in low-energy buildings and thereby gain better indoor lighting conditions. Al-Sallal (1998), a case study for a one-storey house in Fresno, California, also focuses on windows. Here the effects of window size on passive cooling, and passive heating in day lighting are investigated for hot, arid regions. Wall (2006) investigates the first Swedish passive house project, twenty terraced passive houses in

Lindås, Gothenburg. The houses were constructed to meet the target peak power of 10 W/m^2, and were to use supply air heating. The focus of the project was low transmission losses (U=0.16 W/m^2K) and low ventilation losses, which meant using a high-performance heat exchanger (80%) and a high degree of insulation. Special focus was also applied to get the buildings airtight. The average airtightness at 50 Pa was measured to 0.3 $l/s.m^2$, and should be compared to the common praxis in Sweden, 0.8 $l/s.m^2$. In addition to these passive measures a solar heating system was installed to provide about 40% of the heat needed for hot tap water. The paper also include a parametric study of space heating and power demand as a function of set point for heating and cooling, infiltration rate, etc. The simulations are compared to monitored energy performance for the building. The author concludes that the air tightness of the building envelope is essential to meet the targets of $10W/m^2$ and the low heating demand of about 15 $kWh/m^2.a$.

Material selection in passive solar buildings is addressed in Thomas et al. (2006). The paper presents a combination of analytical, experimental and computational studies for selecting affordable materials and designing buildings with the aim of high thermal comfort. The models are validated using measurements in two housing complexes in Egypt.

In Karlsson et al. (2007) the authors make a comparison between three different energy simulation codes, and use a low-energy building as a case. All three models use dynamic models to calculate energy demand for heating and indoor temperature. The low-energy case is a well-known and extensively measured low-energy building in Lindås, Sweden. A parameter of interest in the paper was the small difference between the software's in terms of deviation of energy use. Thus, the paper shows that the relative importance in terms of choice of software is small compared to the large difference in terms of deviation between different households within the studied low-energy buildings. The deviation between software's is shown to be as low as 2%, but the deviation between different households ranges from 6,000 kWh/year to 12,000 kWh/year with an average of 8,020 kWh/year. Furthermore, occupant behaviour, heat exchanger efficiencies as well as air flow control are shown to be important factors. The authors also stress the need for more detailed information about activities and more input data from manufacturers.

In Wang et al. (2009) the authors present a case study of zero-energy house design in the U.K. Zero-energy buildings are defined in the paper as "a building with a net consumption of zero over a typical year." This means that the energy use for heat and electricity is reduced at the same time as this demand is met on an annual basis from renewable energy supply. The renewable can either be building integrated or part of a community renewable energy supply system. A combination of TRANSYS and EnergyPlus is used in the paper, where EnergyPlus is used for building envelope design and TRANSYS for the installations as well as the renewable energy system design. The conclusion of the study is that it is theoretically possible to build zero-energy houses in the U.K. The study also suggests a methodology for the design process where three steps are summarized: (1) analysis of the local climate; (2) application of passive design methods; and (3) investigation of various systems for supply and installations such as PV, wind turbines and solar hot water to optimize the design.

Zhu et al. (2009a) present an energy and economic analysis of a zero-energy house and compare this with a conventional house in Las Vegas. Two houses were built side by side, one zero-energy house and one baseline house, and energy performance measurements were made. The energy contribution from the different components in the building was obtained using Energy10 and eQUEST3.6. The results from these two models gave similar results. The study concludes that four components were clearly economically under given constraints: high-performance windows, compact fluorescent lighting, well insulated roofs and AC units with water-cooled condensers. If financial support was included PV tiles were also considered to have good financial return. Thermal mass walls were found to be too costly. Walls are in focus in Zhu (2009b) where a detailed energy-saving analysis of a high thermal mass wall is presented. This is demonstrated in an actual construction project and compared to a conventional wall. It is shown that for this wall construction the heating use in the building was much lower, but the load slightly higher. According to the study this is due to the effect that more heat is stored during the day than can be returned during the night, increasing the cooling demand. The simulation software used is Energy10 and the experimental part of the study was carried out in Las Vegas.

In Heim et al. (2010) the authors investigate isothermal storage of solar energy in building construction with focus on passive houses. A storage system with phase change materials that absorbs heat during the hot period and releases heat during the cold period is analyzed. The material behaviour is studied using numerical techniques. These methods are then implemented in a general building simulation tool, ESP-r. The paper investigates the effect of a PCM wall and the influence on internal surface temperature. The case is compared with a conventional wall without the PCM material. Both diurnal as well as seasonal latent heat storage is studied. The authors conclude that isothermal heat storage may improve thermal conditions on internal surfaces, but emphasize that the effect of the latent heat storage will depend on its structure, phase change temperature range and total latent heat of the phase change.

4.3 Measurements

Measurements as a method for investigating low-energy buildings is another commonly used method for the papers included in this review. A total of 14 papers are connected to this section. In Dallaire (1980) the concept of zero-energy houses is introduced as a bold low-cost breakthrough that may revolutionize housing. The benefits of super-insulated houses are described in a U.S. and Canadian context, and performance and cost of different components are also described. The background to the development of the concept was the increasing prices of oil in the U.S. The paper includes a series of empirical studies, and among other things the importance of keeping infiltration to a minimum was emphasized.

In Starr et al. (1980) a passive solar house research project is presented. The project demonstrates significant savings in energy. The studied house use 82% less energy than the average California house at the time. The importance of thermal mass, window location and direction as well as insulation is shown. The study contains monitoring for over one year, including both winter and summer conditions.

Nieminen (1994) presents results from the Finnish demonstration houses within IEA Task 13 "Advanced solar low-energy houses". This demonstration project shows an estimated energy use of about 20kWh/m^2, which was below 10% of the average value for small houses in Finland at the time. The main focus of the project was to reduce space heating demand.

Filippin et al. (1998) present the first two years of experiences from a passive solar house in Argentina. The paper presents measurements and the authors conclude that simulation during the design phase had significant advantages, and the internal gains in the form of equipment use patterns have an important influence on the performance.

Schnieders and Hermelink (2006) present the results from the CEPHEUS project, where the material includes measurements and occupant satisfaction for passive houses. The Cost Efficient Passive Houses as European Standard (CEPHEUS) project here includes over 100 dwellings that have been studied. All the projects within the program exhibit extraordinarily low energy use according to the authors, as they can save 80% of space heating compared to ordinary buildings and the total primary energy use was down 50% when compared. It is also concluded that this is achievable with high performance in terms of thermal comfort both in summer and winter for the houses in the project.

In Liu and Henze (2006b) an experimental analysis of simulated learning control for active and passive building thermal storage is reported. In Liu and Henze (2006a) the theoretical foundation is presented and in Liu and Henze (2006b) the results and analysis are found. The work was conducted at the Energy Resource Center Station in Iowa.

The next article discusses how to address the effects of climate change and thermal comfort while at the same time meeting the design challenges of the twenty-first century (Holmes et al., 2007). The authors demonstrate the effect of a changing climate with increasing temperatures using predictions and outline a series of principles in terms of load management, cooling and heating using alternative systems. The paper also shows some of the effects of solar shading as a way of controlling internal gains, as well as the effect of night cooling and other ventilation applications. They also state that the study shows that high-mass buildings are able to provide a higher quality in terms of "internal environment".

Tommerup (2007) presents the results from measurements and discusses how to develop typical single-family houses to meet new energy requirements without compromising on either economy or construction. Tommerup has studied energy-efficient houses built according to the new energy performance requirements in Denmark. The purpose of the project was to demonstrate that it is possible to produce energy-efficient single-family houses that meet existing standards without compromising on economy or construction. Tommerup presents the houses within the project as well as the energy efficiency measures that were applied. A full presentation of energy use, thermal comfort and airtightness is also included. The energy used by these buildings is about 50% to 75% of the typical energy use in Danish buildings in general.

Makaka et al. (2008) present a case study where the building was monitored for a period covering all the South African seasons. The performance of the building was seen to depend mainly on how the occupants used the house. The type of house presented was shown to represent a lower rate of temperature and humidity variations. The thermal behaviour and ventilation efficiency of a low-cost passive solar energy efficient house is investigated. The low-cost houses in South Africa are categorized by poor craftsmanship in terms of energy-efficient design and passive solar features, thermal climate and ventilation efficiency. If this type of design is used, large savings are possible.

Chandel and Aggarwal (2008) have done an evaluation of the performance of a passive solar building. The building is located in the Western Himalayas. The heat losses were shown to be reduced by approximately 35% with passive solar measures. In Wojdyga (2009) the author investigates the heat demand in a low-energy building in Poland. Results from a five-year study of the energy consumption in a single-storey terraced low-energy house are presented.

Maier et al. (2009) combines methods when presenting a comparison of physical performance of a ventilation system in residential low-energy buildings. To analyze the influence of ventilation systems on comfort, the authors used a combination of measurements and a questionnaire. The measurement part of the project included 22 residential houses in Germany, chosen and equipped with four different types of ventilation systems: (1) natural ventilation; (2) air heating system; (3) mechanical ventilation with supply and exhaust with heat exchanging; and (4) mechanical ventilation with single ventilators. The monitor's parameters were CO_2, relative humidity, air temperature, electricity, gas and heat. The use of the window openings, use of ventilation and number of residents present were also presented. The mechanical ventilation performed better in terms of CO_2 concentration than the naturally ventilated cases.

Feist and Schnieders (2009) explain the concept of the passive house technique and issues such as design methods, components in a passive house such as thermal bridges, windows, junctions of roof and wall, etc., the importance of internal gains and ventilation and air tightness issues. Practical experience is also summarized. In a similar way Nicoletti (1998) discusses low-energy design from an architectural point of view. Form as a tool for energy control is discussed, and several examples and objects such as the University building in Udine, Casa Moncada and a headquater of a bank in Rome. A special discussion for tall buildings is included.

Kalz et al. (2010) have also done a long-term study. The authors present what they define as a holistic approach when evaluating heating and cooling using building signatures. The study includes a comprehensive analysis of eleven low-energy buildings in terms of energy use and thermal comfort. The long-term study is presented using detailed time series for between two and five years depending on object. In the paper a methodology is described for evaluating heating and cooling concepts, not only by focusing on thermal comfort but also by including the useful energy consumption and energy efficiency, generation, distribution and deliveries.

4.4 Interviews

Two papers are related to this section, one interdisciplinary study of the low-energy houses in Lindås, Sweden and one article investigating the attitude of large construction companies.

Isaksson and Karlsson (2006) present an interdisciplinary study of the indoor climate in the low-energy buildings in Lindås in Sweden. The paper presents results from an investigation of the 20 terraced houses in Gothenburg in Sweden. Qualitative interviews with occupants are combined with physical measurements of the thermal environment. The results show that when occupants are present and appliances are used, the temperature can be managed within acceptable limits even during cold days. One main outcome of the study is the importance of information given to households about the functionality of the heating system. In addition to this the authors state that temperature control could be improved. The paper also gives a wider view of how activities within these houses change compared to normal houses where power is less of a problem. Special focus should also be placed on gable houses in terms of thermal comfort.

In Hamza and Greenwood (2009) the impact of the new energy conservation regulations and its impact on low-energy buildings is studied. Data collection was made by semi-structured interviews with a sample representing large construction companies, architectural practitioners and building performance consultants. The authors express that "overall, it appears likely that the legislation is already having a profound effect on the contractual and procurement arrangements of U.K. construction projects." In Hamza and Greenwood (2009) a number of interesting impacts of the new legislation is seen on: (1) tendering practice and documentation; (2) procurement practice; (3) post-tender engineering and "value engineering"; and (4) collaborative working.

4.5 Questionnaires

Only one article fell under this category. In Thomsen et al. (2005) twelve demonstration projects within IEI Task 13 "advanced solar low-energy buildings" are presented. The paper includes a brief presentation of the houses. The study is a follow-up study three years after Task 13 ended, and is made in the form of a questionnaire sent to the former participants within the task. The paper states that the measured energy use was in general higher than expected due mainly to unforeseen technical problems but that energy savings of 60% were achieved compared to a typical building. The question of overheating in summer was specifically addressed, and it was shown that with proper planning and design this can be avoided. However, within the project this was a problem in two cases, one in Norway and one in Denmark. The paper summarizes a series of lessons to be learned: (1) Special consideration should be given to heat losses in partitions between apartments in highly insulated buildings; (2) obtaining the needed air tightness of a house requires careful planning and control of seals and barriers; (3) ventilation should be designed carefully with regards to sound and draught; (4) overheating can be prevented in moderate climates by means of thermal mass, solar shading and ventilation, if they are designed properly; and (5) heat losses from ducts and pipes are important and should be minimized.

4.6 Environmental or life cycle focus studies

In this final section nine papers are reviewed. In Chwieduk (1999) a study of thermal modernisation and refurbishment of existing buildings are presented in a Polish context. The paper outlines a view that a transition to low-energy buildings in Poland is a natural progression. It also includes some remarks and recommendations for Polish low-energy buildings.

Tombazis and Preuss (2001) discuss design of solar buildings in an urban context. The study emphasizes the building's access to natural resources while taking into account the negative influence that may prevail around the site. The associated constraints, according to the authors, are challenging but very interesting and rewarding from an architectural point of view. The paper exemplifies different design options for three different cases. Even though the buildings are different they share some features, such as: (1) well insulated; (2) shallow plan, so daylight is able to penetrate and also to achieve well functioning natural ventilation; and (3) hybrid ventilation systems are used and some form of intelligent control is included.

Zimmermann et al. (2005) present a benchmark of sustainable construction. The paper addresses the policy field and has the aim of being a contribution to developing a standard in the field. The paper also shows that buildings designed to the passive house standard may comply with the requirements for sustainable construction if the electricity generation is based on environmentally friendly generation. However, for other parts where a high degree of fossil fuel is used the authors find it much harder to meet the requirements.

In Rabah (2005) a design strategy for energy-efficient passive solar buildings in Cyprus is presented. The methodology includes: (1) initial pre-design considerations; (2) initial climate analysis; (3) determination of passive solar design strategies; (4) analysis of the control zone (comfort, etc.). The aim of the paper to provide general information at pre-design phases to be able to more effectively implement passive solar energy. Krishan et al. (1996) also focus on the design phase and discuss climate responsible design for two cases, one high-altitude "cold-dry" case and one "hot and dry" case. The paper includes design principles for this "indigenous architecture of two Indian deserts".

In Thormark (2006) the effect of choice of material on total energy use and recycling potential is reported. The author addresses both the need for reducing energy use as well as the maximization of the recycling potential. Since the embodied energy for a low-energy house accounts for a large part of the total energy use during the life span of the building, it is important to consider the choice of material. The article presents the impact of material choice on the passive houses built in Lindås, Sweden.

In Sartori and Hestnes (2007) energy use in conventional building is compared to low-energy buildings using a review approach. The review includes 60 cases from nine countries, and showed that by far the largest part of energy use is related to the operating phase. The results presented in the paper show that the solar houses proved to be more energy efficient than the houses within the studies that used "green" materials. Furthermore, it was shown that solar houses decreased life-cycle energy use by half compared to a conventional building. A passive house proved to be more efficient than the solar houses in the studies.

In Aste et al. (2010) the low-energy residential settlement in Borgo Solare, Italy is presented. The project is not just an experimental operation; instead Borgo Solare is a real urban district. However, the project may be considered to be an advanced and innovative residential area designed on sustainable architectural grounds. The paper presents a techno-economical analysis of the project. The analysis shows that the higher initial embodied energy in a low-energy building may be paid back well within the life span of the building. In the economical analysis the authors argue that the higher initial costs may be effective in the long term.

In Verbeeck and Hens (2010) a life-cycle inventory of buildings is presented. The paper presents results of a contribution analysis of the life-cycle inventory of four typical buildings. The location of the objects is Belgium. The paper shows the small importance of the embodied energy when comparing energy use during the buildings' entire usage phase. This is also shown to be even more so for energy efficiency measures, when comparing embodied energy of the measures with the reduction in use. Only extreme low-energy buildings may have a higher embodied energy than the energy use during the phase when it is used; for a normal building this represents about one third of the total energy use during the life cycle. The total savings, however, are still shown to be large for low-energy dwellings.

5. Concluding discussion

The attention and research in this field is characterized by a strong increase in the number of articles during the last five years, not least within the scope of low-energy buildings and solar buildings. However, it is also important to note that most of the development in the field is taking place outside the scientific community, in construction companies, national programs and housing companies. The research field, as presented here, is a clearly technical field with a strong focus on the technologies and development of techniques for improving energy performance of buildings. The number of studies with focus on the end users and how they interpret and interact with this new technology is scarce, but there are examples, such as Isaksson and Karlsson (2006), Schnieders and Hermelink (2006) and Feist et al. (2005), to name a few important contributions. This is a field which is important, especially when several authors, among them Karlsson et al. (2007), stress the importance of the activities and internal gains for low-energy buildings. The relative importance of this factor is so much greater since the losses from the building are so much smaller. It is therefore even more important to understand and be able to predict activities when designing this type of building.

The general trend of publication can also be said to have started to shift if looking at the process from 1978 to the present day. In the late 1970s and early 1980s the focus in the presented articles was in general on single technology investigations or building oriented with energy use and cost as key focus. The main driver was to remove oil-fired burners or to minimize their use, as an effect of the oil crisis. A shift can be seen in this main focus, as a large part of the articles presented in the late 1990s and after 2000 have a more environmental focus with greenhouse emissions as a key focus. This is in line with the general trend. However, what may be of interest is the increasing number of policy articles that argue for low-energy buildings when looking at long-term scenarios for sustainable buildings or regions. The number of publications where sustainable city parts like Borgo Solare in Italy are reported is also increasing. So the general trend may be argued to be moving from single technology and individual case studies of buildings to a more regional and large-scale production of energy-efficient city parts. In Lindås a in comparison small-scale production of 20 terraced passive houses was constructed in Sweden. These houses are investigated from multiple perspectives and some of the material is reported in the references here, such as Karlsson and Moshfegh (2006) and Wall (2006) for a technical description of the implementation and Isaksson and Karlsson (2005) for an interdisciplinary study of the buildings and also Thormark (2006) for a study of embodied energy and life-cycle analysis of these buildings.

For Sweden the buildings in Lindås are important as they mark the starting point for building passive houses in Sweden. Due to that they also represent a starting point in terms of learning to build this type of building with low infiltration rates and a high level of insulation, etc. which requires different approaches from the construction industry in terms of process. One interesting factor is that the German standard for passive houses sees e.g. Fiest and Schnieders (2009) or Fiest et al. (2005), has been adapted to Swedish conditions by a national forum for energy-efficient buildings funded by the Swedish energy agency (FEBY). This trend is similar for several other European countries. One point of interest is how the national standards use the requirements within the German standard in cases such as for Sweden, where the climate is different. For Sweden the certification process has the same requirements on maximum power and energy. These are based on electric heating, using the indoor air quality designed airflow for the building (10 W/m2) and a maximum energy use of 15 kWh/m^2. This is of course harder to achieve in a Nordic climate than for a central European climate. This issue is also connected to the issue of thermal comfort in passive houses in Nordic regions, where relatively few studies are reported. However, the issue of thermal comfort in general is something that is becoming more common to investigate, see Feist (2009) for example, but further studies are still very much needed especially for cold climates.

Along with users' interaction and interpretation of low-energy buildings, user satisfaction was expected to be a main focus of articles in this compilation. However, only one (Isaksson and

Karlsson 2005) explicitly tried to explore this. There are no internationally standardized methods to evaluate user satisfaction, but a closed-end questionnaire on indoor climate in dwellings has been developed in, for example, Sweden (Andersson et al., 1988). However, research results from low-energy buildings in Web of Science using this questionnaire are lacking. Also in Germany, questionnaires have been used in research about user satisfaction, but in office buildings (Pfafferott et al., 2007; Wagner et al., 2007). The method has been developed by University of California's Center for Environmental Design Research, Berkeley and according to the authors it addresses "all relevant aspects of occupant satisfaction with indoor environments" (Wagner et al., p. 764). Results show how user satisfaction corresponds to control abilities for users which are supported by results in Pfafferott et al. (2007). Actual temperature and temperature sensations had less effect on user satisfaction in this study. The perceived flexibility of low-energy buildings is something that future research could address. Post-occupancy evaluations of office buildings might offer methodological inspiration.

Research focusing on the construction sector, clients, design teams and the organization of construction processes are in this compilation mainly found in the U.K. (cf. Hamza and Greenwood, 2009; Hamza and Horne, 2007). Although the articles analyse phenomena specific to the U.K. (new energy conservation regulations and low-energy architecture in higher education), some general conclusions can be made. When designing low-energy buildings, more relational thinking is needed because of the increased complexity in the design phase (Hamza and Horne, 2007). Students in architecture might not have sufficient training in this higher level of approaching tasks, which includes critical thinking. Modules are being developed, however, to incorporate and facilitate relational thinking. A new regulation on energy conservation in the U.K. has also proved to support collaborations between design and construction teams, which is considered most welcome (Hamza and Greenwood, 2009). As noted in Hamza and Greenwood (2009), it is important not only to study user satisfaction post occupancy, but also the experiences of design and construction teams, in order to improve present regulations and practice in construction processes. Groups that should be addressed are practitioners, educators and policy-makers and publications in Web of Science journals might not be the most effective way to disseminate this feedback.

6. References

Al-Sallal, KA. (1998). Sizing windows to achieve passive cooling, passive heating, and daylighting in hot arid regions, In: *Renewable energy*, 14 (1-4): MAY-AUG 365-371

Andersson, K.; Fagerlund, I.; Bodin, L.; Ydreborg, B. (1988). Questionnaire as an instrument when evaluating indoor climate. In: *Healthy Buildings´88 Stockholm* 1988, Vol 1:139-146

Aste, N.; Adhikari, RS.; Buzzetti, M. (2010). Beyond the EPBD: The low energy residential settlement Borgo Solare, In: *Applied Energy*, 87 (2): FEB 629-642

Babbie, E. (1990) Survey research methods (2nd ed.),: Wadsworth. 0-534-12672-3 Belmont CA

Badescu, V. (2005). Simulation analysis for the active solar heating system of a passive house, In: *Applied Thermal Engineering*, 25 (17-18): DEC 2754-2763

Badescu, V.; Sicre, B. (2003). In: Renewable energy for passive house heating II. Model, In: *Energy and Buildings*, 35 (11): DEC 1085-1096

Badescu, V.; Sicre, B. (2003). Renewable energy for passive house heating Part I. Building description, In: *Energy and Buildings*, 35 (11): DEC 1077-1084

Bergsten, B. (2001) Energiberäkningsprogram för byggnader – en jämförelse utifrån funktions- och användaraspekter, Effektivrapport.
Bergström, G.; Boréus, K. (2005). Textens mening och makt: metodbok i samhällsvetenskaplig text- och diskursanalys. (2., [omarb.] uppl.) Studentlitteratur. ISBN: 91-44-04274-4, Lund
Chandel, S.; Aggarwal, R. (2008) Performance evaluation of a passive solar building in Western Himalayas, In: *Renewable energy*, 33 (10): OCT 2166-2173
Chlela, F.; Husaunndee, A .; Inard, C.; RiedeFer, P. (2009) A new methodology for the design of low energy buildings, In: *Energy and Buildings*, 41 (9): SEP 982-990
Chwieduk, D. (1999). Prospects for low energy buildings in Poland, In: *Renewable energy*, 16 (1-4): JAN-APR 1196-1199
Clarke, J.; Grant, A.; Johnstone, C.; Macdonald, I. (1998). Integrated modelling of low energy buildings, *Renewable energy*, 15 (1-4): SEP-DEC 151-156
Cooper, J.S.; Fava, J. (2006). Life Cycle Assessment Practitioner Survey: Summary of Results, In: *Journal of Industrial Ecology*, 10(4) 12 -14
Crawley, D. Hand, J. Kummert, M. Griffith, B. (2005) Contrasting the capabilities of building energy performance simulation programs. In *proceedings of international IBPSA conference 8*, Montreal, Canada, 231-238.
Dallaire, G. (1980). Zero-energy house: bold, low-cost breakthrough that may revolutionize housing, In: *Civil Engineering –ASCE*, 52, 47-59
Dinçer, I.; Rosen, M.A. (2007). *Exergy: energy, environment and sustainable development*. (1. ed.), Elsevier, ISBN: 978-0-08-044529-8, Oxford
Fairclough, N. (2003). Analysing discourse: textual analysis for social research. Routledge, ISBN: 0-415-25893-6, New York
Feist, W.; Schnieders, J. (2009). Energy efficiency - a key to sustainable housing, In: *European Physical Journal – Special Topics*, 176: SEP 141-153
Feist, W.; Schnieders, J.; Dorer, V.; Haas, A. (2005) Re-inventing air heating: Convenient and comfortable within the frame of the Passive House concept, In: *Energy and Buildings*, 37 (11): NOV 1186-1203
Filippin, C.; Beascochea, A.; Esteves, A.; De Rosa, C.; Cortegoso, L.; Estelrich, D. (1998) A passive solar building for ecological research in Argentina: The first two years experience, In: *Solar Energy*, 63 (2): AUG 105-115
Frosch, RA.; Nicholas EG. (1989). Strategies for Manufacturing, In: *Scientific American* 261(3): 144-152
Guinée, JB. (red.) (2002). Handbook on life cycle assessment: operational guide to the ISO standards, Kluwer, ISBN: 1-4020-0228-9, Dordrecht
Hamza, N.; Greenwood, D. (2009) Energy conservation regulations: Impacts on design and procurement of low energy buildings, In: *Building and Environment*, , 44 (5): MAY 929-936
Hamza, N.; Horne, M. (2007). Educating the designer: An operational model for visualizing low-energy architecture, In: *Building and Environment*, 42 (11): NOV 3841-3847
Heim, D. (2010) Isothermal storage of solar energy in building construction, In: *Renewable energy*, 35 (4): APR 788-796
Holm, D. (1996). The status of low energy architecture in South Africa, In: *Renewable energy*, 8 (1-4): MAY-AUG 301-304
Holmes, M.; Hacker, J. (2007). Climate change, thermal comfort and energy: Meeting the design challenges of the 21st century, In: *Energy and Buildings*, 39 (7): 802-814

International Organization for Standardization (ISO) (1997). *Environmental management - Life cycle assessment - Principles and framework.* Geneva: ISO

IPCC (ed. Terry Baker) (2007). *Climate Change 2007: Synthesis Report*, http://www.ipcc.ch/ipccreports/ar4-syr.htm (April 21 2009).

Isaksson, C.; Karlsson, F. (2006). Indoor climate in low-energy houses - an interdisciplinary investigation, In: *Building and Environment*, 41 (12): DEC 1678-1690

Kalogirou, SA.; Bojic, M. (2000). Artificial neural networks for the prediction of the energy consumption of a passive solar building, In: *Energy*, 25 (5): MAY 479-491

Kalz, D.; Pfafferott, J.; Herkel, S. (2010). Building signatures: A holistic approach to the evaluation of heating and cooling concepts, In: *Building and Environment*, 45 (3): MAR 632-646

Karlsson, F.; Rohdin, P.; Persson, M-L. (2007). Measured and predicted energy demand of a low energy building: Important aspects when using building energy simulation, In: *Building Services Engineering Research and Technology*, 28 (3): 223-235

Karlsson, J.; Moshfegh, B. (2006) Energy demand and indoor climate in a low energy building-changed control strategies and boundary conditions, In: *Energy and Buildings*, 38 (4): APR 315-326

Krishan, A.; Jain, K.; Tewari, P. (1996). Indigenous architecture of two Indian deserts and modern climatic responsive solutions, In: Renewable energy, 8 (1-4): MAY-AUG 272-277

Krosnick, JA. (1999). Survey Research, In: *Annual Review of Psychology*, 50: 537-567

Kvale, S (1996). *Interviews An Introduction to Qualitative Research Interviewing*, Sage Publications, Thousand Oaks.

Kvale, S.; Brinkmann, S. (2009). *InterViews: learning the craft of qualitative research interviewing*. (2nd ed.) Sage Publications, ISBN: 978-0-7619-2542-2, Los Angeles

Körner, S.; Ek, L.; Berg, S. (1984). *Deskriptiv statistik*. (2. ed.) Studentlitt., ISBN: 91-44-15392-9, Lund.

Lincoln, Y.S.; Guba, EG. (1985). *Naturalistic inquiry*. Beverly Hills, Calif.: Sage.

Liu, S.; Henze, G. (2006a). Experimental analysis of simulated reinforcement learning control for active and passive building thermal storage inventory Part 1. Theoretical foundation, In: *Energy and Buildings*, 38 (2): FEB 142-147

Liu, S.; Henze, G. (2006b). Experimental analysis of simulated reinforcement learning control for active and passive building thermal storage inventory Part 2: Results and analysis, In: *Energy and Buildings*, 38 (2): FEB 148-161

Lomas, K. (1996). The UK Applicability Study: An evaluation of thermal simulation programs for passive solar house design, In: *Building and Environment*, 31 (3): MAY 197-206

Maier, T.; Krzaczek, M.; Tejchman, J. (2009). Comparison of physical performances of the ventilation systems in low-energy residential houses, In: *Energy and Buildings*, 41 (3): MAR 337-353

Makaka, G Meyer, E McPherson, M, Thermal behaviour and ventilation efficiency of a low-cost passive solar energy efficient house, In: *Renewable energy*, 33 (9): 1959-1973 SEP 2008

Moses, LE. (1986). *Think and explain with statistics*. Addison-Wesley ISBN 0-201-15619-9, Reading, Mass.

Nicoletti, M. (1998). Architectural expression and low energy design, In: *Renewable energy*, 15 (1-4): SEP-DEC 32-41

Nieminen, J. (1994). Low-energy residential housing, In: *Energy and Buildings*, 21 (3): 187-197

Odum, H.T. (2007). *Environment, power and society for the twenty-first century: the hierarchy of energy.* (New ed.) University Press. ISBN: 978-0-231-12886-5 New York: Columbia

Onishi, J.; Soeda, H.; Mizuno, P. (2001). Numerical study on a low energy architecture based upon distributed heat storage system, In: *Renewable energy,* 22 (1-3): JAN-MAR 61-66

Osgood, CE.; Suci, GJ.; Tannenbaum, P. (1957). *The Measurement of Meaning,* University of Illinois Press, Urbana, IL

Persson, M-L.; Roos A.; Wall, M. (2006). Influence of window size on the energy balance of low energy houses, In: *Energy and Buildings,* 38 (3): MAR 181-188

Pfafferott, JÜ.; Herkel, S.;. Kalz, DE.; Zeuschner, A. (2007) Comparison of low-energy office buildings in summer using different thermal comfort criteria, In: *Energy and Buildings,* 39 pp. 750-757

Rabah, K. (2005). Development of energy-efficient passive solar building design in Nicosia Cyprus, In: *Renewable energy,* 30 (6): MAY 937-956

Sartori, I.; Hestnes, A. (2007) Energy use in the life cycle of conventional and low-energy buildings: A review article, In: *Energy and Buildings,* 39 (3): MAR 249-257

Schnieders, E.; Hermelink, A. (2006). CEPHEUS results: measurements and occupants' satisfaction provide evidence for Passive Houses being an option for sustainable building, In: *Energy Policy,* 34 (2): JAN 151-171

Shaviv, E.; Yezioro, A.; Capeluto, IG.; Peleg, UJ.; Kalay, YE. (1996). Simulations and knowledge-based computer-aided architectural design (CAAD) systems for passive and low energy architecture, In: *Energy and Buildings,* 23 (3): MAR 257-269

Starr, G.; Neubauer, L.; Melzer, B. (1980). Temperature Constrol by Passive Solar House Design in Calefonia, In: *Transactions of the ASAE,* , 23 (2): 449-456

Thomas, J.; Algohary, S.; Hammad, F.; Soboyejo, W. (2006). Materials selection for thermal comfort in passive solar buildings, In: *Journal of Materials Science,* 41 (21): NOV 6897-6907

Thomsen, K.; Schultz, J.; Poel, B. (2005). Measured performance of 12 demonstration projects - IEA Task 13 "advanced solar low energy buildings", In: *Energy and Buildings,* 37(2): FEB 111-119

Thormark, C. (2006). The effect of material choice on the total energy need and recycling potential of a building, In: *Building and Environment,* 41 (8): AUG 1019-1026

Tombazis, A.; Preuss, S. (2001). Design of passive solar buildings in urban areas, In: *Solar Energy,* 70 (3): 311-318

Tommerup, H.; Rose, J.; Svendsen, S. (2007). Energy-efficient houses built according to the energy performance requirements introduced in Denmark in 2006, In: *Energy and Buildings,* 39 (10): OCT 1123-1130

Verbeeck, G.; Hens, H. (2010). Life cycle inventory of buildings: A contribution analysis, In: *Building and Environment,* 45 (4): APR 964-967

Versteeg, H. K. Malalasekera, W. (1995) *An introduction to computational fluid dynamics: the finite volume method.* Harlow: Longman Scientific & Technical.

Visser, PS.; Krosnick, JA.; Marquette, J.; Curtin M. (1996). Mail surveys for election forecasting? An evaluation of the Columbus Dispatch poll. In: *Public Opin. Q.* 60:181.227

Wagner, A.; Gossauer, E.; Moosmann, C.;. Gropp, Th.; Leonhart, R. (2007) Thermal comfort and workplace occupant satisfaction—Results of field studies in German low energy office buildings, In: *Energy and Buildings* , 39, pp. 758-769

Wall, M. (2006). Energy-efficient terrace houses in Sweden - Simulations and measurements, In: *Energy and Buildings,* 38 (6): JUN 627-634

Wang, L.; Gwilliam, J.; Jones, P. (2009) Case study of zero energy house design in UK, In: *Energy and Buildings*, 41 (11): NOV 1215-1222

Wojdyga, K. (2009). An investigation into the heat consumption in a low-energy building, In: *Renewable energy*, 34 (12): DEC 2935-2939

Zhu, L.; Hurt, R.; Correa, D.; Boehm, R. (2009a). Comprehensive energy and economic analyses on a zero energy house versus a conventional house, In: *Energy*, 34 (9): SEP 1043-1053

Zhu, L.; Hurt, R.; Correia, D.; Boehm, R. (2009b). Detailed energy saving performance analyses on thermal mass walls demonstrated in a zero energy house, In: *Energy and Buildings*, 41 (3): MAR 303-310

Zimmermann, M.; Althaus, H.; Haas, A. (2005). Benchmarks for sustainable construction - A contribution to develop a standard, In: *Energy and Buildings*, 37 (11): NOV 1147-1157

7

Energy transformed: building capacity in the engineering profession in Australia

Cheryl Desha and Karlson 'Charlie' Hargroves
Griffith University, Curtin University
Australia

1. Introduction

Global pressures of burgeoning population growth and consumption are threatening efforts to reduce negative environmental pressures associated with development such as atmospheric, land and water pollution. For example, the world's population is now growing at over 70 million per year or 1 billion per decade (Brown, 2007), increasing from 3.5 billion in 1970, to 5 billion in 1990, to 7 billion by 2010 (United Nations, 2002). In 1990 only 13 percent of the global population lived in cities, while in 2007 more than half did. More than 60 percent of the global population lives within 100 kilometers of the coastline (World Resources Institute, 2005) and nearly all of the population growth hereon is forecast to happen in developing countries (Postel, 1999). Future levels of stress on the global environment are therefore likely to increase if current trends are used for forecasting, which is particularly challenging as scientists are already observing significant signs of degradation and failure in environmental systems. For example, the Intergovernmental Panel on Climate Change *Fourth Assessment Report* (IPCC, 2007) provided an unequivocal link between climate change and current human activities, in particular: the burning of fossil fuels; deforestation and land clearing; the use of synthetic greenhouse gases; and decomposition of wastes from landfill. The UK *Stern Review* concluded that within our lifetime there is between a 77 to 99 percent chance (depending on the climate model used) of the global average temperature rising by more than 2 degrees Celsius (Stern, 2006), with a likely greenhouse gas concentration in the atmosphere of 550 parts per million (ppm) or more by around 2100.

Hence, the way in which the human race deals with energy over the next 30 years will determine the quality of life for generations to come. This includes increasing the level of non-fossil fuel energy generation, improving the efficiency with which energy is supplied (i.e. reducing supply losses, for example from the power station to the end user), and increasing the efficiency of the end user (i.e. reducing demand, for example improving the efficiency of machinery and appliances to perform, for example heating, cooling, and washing). However, the reality is that there are significant skills shortages - particularly within the engineering profession - to address these issues. As highlighted in a United Nations Environment Program report on working in a low-carbon world, '... *companies in the fledgling green economy are struggling to find workers with the skills needed to perform the work that needs to be done. Indeed, there are signs that shortages of skilled labor could put the brakes on green expansion…There is thus a need to put appropriate education and training arrangements in place'* (United Nations Environment Programme, 2008).

In Australia for example, considering energy efficiency, according to a national study, 'Given the wide range of technical issues associated with energy efficiency, gaps in the skill sets of specialists such as engineers or trades people could prevent the uptake of these options across a range of sectors' (Garnaut, 2008). Modelling by the CSIRO has shown that 3 million Australians (of a population of just over 20 million), will need training or re-training in energy efficiency, green building technologies, sustainable energy and more sustainable agricultural systems to enable Australia to achieve the IPCC's recommended targets for greenhouse gas reductions (Hatfield-Dodds et al., 2008). Furthermore, surveys are highlighting that the state of knowledge, understanding and implementation of even basic environmental and energy management systems in the business sector is poor. For instance, a 2008 survey of 300 Australian business CEOs regarding operating in a carbon-constrained economy found that two-thirds (67 percent) of businesses were concerned or unsure about compliance obligations, and only a handful of businesses (less than 3 percent) had implemented a strategic response to climate change (Price Waterhouse Coopers, 2008). A 2007 survey of the Australian mining and metals sector also highlighted an alarmingly slow adoption of energy demand management practices, with nearly half (43 percent) of companies still not having implemented an official energy policy. In the same context, only 10 percent of companies responding to a 2007 national Australian Industry Group survey on climate change practices felt informed enough to manage the risks associated with climate-related impacts (Australian Industry Group, 2007). Australia's peak engineering professional body, the Institution of Engineers Australia, has also acknowledged that, 'The need to make changes in the way energy is used and supplied throughout the world represents the greatest challenge to engineers in moving toward sustainability' (Institution of Engineers Australia, undated).

Within this context, this chapter overviews the need to transform engineering education, to deliver graduates and capacity build professionals who can address such energy supply and demand challenges, highlighting the complexity of transforming such education systems. We begin by discussing the problem of a time lag dilemma facing education worldwide, whereby the timeframe for capacity building the profession is converging with the requirement for global action. We also briefly discuss related risks and benefits facing organizations in light of this dilemma. The chapter then focuses on requirements for capacity building the engineering profession, drawing on two research initiatives that we have previously led: a 2007 Australian survey of the state of energy efficiency education in engineering education, and a 2009 investigation into increasing the extent of energy efficiency content in curriculum. Finally, we discuss a peer reviewed, online and freely accessible resource that has been developed from this increased understanding, to assist with capacity building, focusing on sustainable energy solutions for climate change mitigation.

2. A time lag facing engineering education

With considerable advances over the last century, the effectiveness of the education system to deliver skilled professionals would appear to be self-evident. Yet signals now clearly suggest that the focus of much of higher education requires a significant update, with an emerging time related imperative facing the engineering community. Despite an absence of discussion in the literature, anecdotal evidence from discussions with engineering educators suggests that a typical (or 'standard') process of curriculum renewal in the higher education sector may take 3-4 accreditation cycles (of approximately 5-year intervals) for engineering departments to fully integrate a substantial new set of knowledge and skills within all year levels of a degree as required; i.e. between 15-20 years.

Given that the average pathway to graduate from an engineering and built environment program is approximately 3-5 years, from enrolment to graduation, followed by 3-5 years of on-the-job graduate development, if HEIs take the typical approach over a 15-20 year period to fully renew such bachelor programs, this has the potential to result in a time lag of around 21-29 years - 2-3 decades - before students graduating from fully integrated programs will be in decision-making positions. Clearly this is well beyond the timeframes needed to address immediate climate change issues. For postgraduate students the time lag may be shorter as students may already be practising in their field and studies span just 1-2 years. However the time lag may still be in the order of 5-10 years depending on the curriculum renewal process, which still potentially results in a lack of capacity in the professional sector over the next decade to address urgent climate change and sustainable development issues.

Along with understanding that current education systems are yet to be prepared to rapidly develop knowledge and skills related to reducing environmental pressures, it is important to understand that it is logistically impossible for the education system to change 'overnight', as programs need to balance the current student demands and expectations with industry expectations for graduate attributes. Figure 1 highlights how the transition might occur, with a period of rapid curriculum renewal followed by continual program improvement which follows a regular improvement cycle of research, curriculum development, trial, evaluation and review.

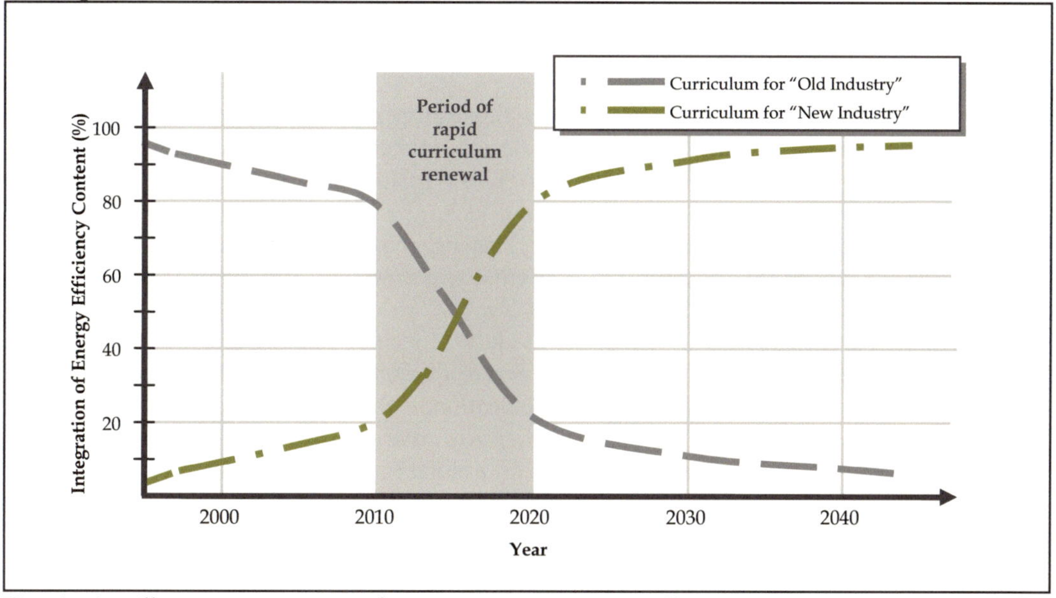

Fig. 1. An illustrative scenario for integrating energy efficiency knowledge and skills to match industry requirements over time

A key consideration in timing the transition, is the shift in focus from 'old industry' to 'new industry' curriculum, matching changing educational needs with the pace of emerging demand for such graduate attributes by employers. As part of the transition towards more sustainable infrastructure and societies, 'old industry' plant and equipment will require service and maintenance by professionals with 'old industry' knowledge and skills. However as with any major adjustment such as the information technology revolution, there needs to be a staged approach, where the balance of 'old' and 'new' needs to be carefully managed in relation to the emerging needs of society and employer demands. As the large amount of embedded infrastructure (for example buildings, power stations, electricity grids etc) needs to be managed, maintained and transitioned, this requires 'old industry' education. Hence the process to integrate 'new industry' knowledge and skills needs to be appropriately staged, as if it is too quick, this could be problematic as graduates may not have the skills that the employment market needs at the time that they graduate.

Hence, the timeframe for updating undergraduate engineering curriculum using standard methods may be too long to ensure that engineering professionals will be equipped with knowledge and skills that can address such immediate 21st Century challenges while still being able to maintain current systems. The extent of the time lag will depend on how quickly the new knowledge and skills are embedded into engineering curriculum, to the point where a student can begin studies in first year, and fully develop the new set of desired knowledge and skills (or 'graduate attributes') by the time they graduate.

This observed time lag dilemma facing engineering education has significant implications for society if the need for curriculum renewal is not addressed. Furthermore, there are implications for university engineering departments as they make decisions about the scale and pace of curriculum renewal as regulations and the market continue to change. Engineering departments may also be exposed to potential risks with regard to both student demand for the programs, and tightening accreditation requirements. However, departments need to be wary of keeping pace with graduate demand (i.e. not stepping too far in front) to ensure that their graduates remain employable and in demand throughout the process.

Drawing on the literature, Figure 2 presents an illustrative representation of the relationship between a department's commitment to engineering education for sustainable development and potential risk and reward implications. Risks include for example falling student numbers, increasing accreditation difficulties, poaching of key staff. Rewards include for example attracting the best students and staff, staying ahead of accreditation requirements, attracting research funding, securing key academic appointments and industry funding.

For the last 20 years, there has been relatively low risks and benefits from seeking to accelerate curriculum renewal in this area, evidenced by the relative lack of action on the whole in the sector apart from a small number of outstanding cases (Desha et al., 2009). However, recent market, regulatory and institutional shifts around environmental and sustainable development related issues, together with the significant shift in public opinion on these matters, and the increasing competition among higher education institutions, have caused the level of both the risks and the benefits to increase dramatically over the coming decades.

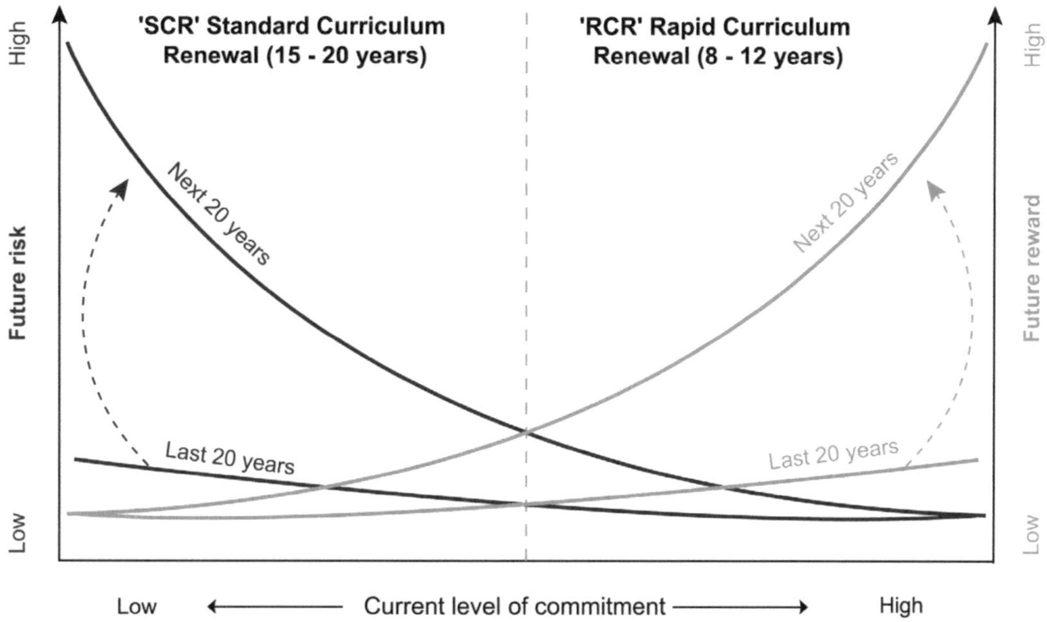

Fig. 2. A stylistic representation of risk and reward scenarios for curriculum renewal in the higher education sector
Source: (Desha & Hargroves, 2009a)

This situation presents significant cause for universities and engineering departments to rethink their strategies related to curriculum reform in order to minimize the risks and capture the rewards. In short, over the coming years, departments who do not transition their programs with topic areas such as energy efficiency are likely to find it increasingly difficult to operate. Furthermore, their traditional roles as providers of education for engineers may be challenged by private training providers who explore niche business opportunities in capacity building in these topic areas, along with engineering firms and government departments developing in-house capacity building programs that assume a base-line graduate capacity.

3. The state of energy efficiency education

In the face of such a time lag dilemma, the literature suggests that engineering educators need to undertake rapid curriculum renewal to update what is taught, within 2-3 accreditation cycles in undergraduate programs. Furthermore, rapid curriculum renewal in postgraduate engineering education also needs to occur; equipping practitioners and decision-makers with knowledge and skills surround energy efficiency. With this in mind, we now consider the state of engineering education for energy efficiency, for which a full account is provided by Desha et al. (2007). We also identify challenges and opportunities for energy efficiency education within universities, for which a full literature review is available online (Desha, Hargroves & Reeve, 2009) and a summary is provided by Desha and Hargroves (2009b).

3.1 Understanding the state of engineering education for energy efficiency

The sub-topic of energy efficiency is a prime example for a new area of practice that needs to be rapidly integrated into engineering courses, while also addressing a knowledge gap in a highly topical content area. However, there is an absence of literature documenting the state of affairs, to provide a robust platform on which to act. Hence, in 2007 the National Framework for Energy Efficiency (NFEE) funded researchers from Griffith University (The Natural Edge Project, TNEP) to undertake the first survey of energy efficiency education across all Australian universities teaching engineering education, which asked, '*What is the state of education for energy efficiency in Australian engineering education?*' (Desha et al., 2007).

The subsequent research project used a paper-based questionnaire which was issued in hard copy and electronic format to the heads of department of all 32 Australian universities providing engineering undergraduate and/or post-graduate programs. It included an invitation to every Dean for completion by every lecturer teaching energy related material within engineering education. The project also included a student questionnaire, which was provided to all lecturers who received the lecturer questionnaire, to distribute and collect in one or more of their classes where energy related material were taught. The results of the two questionnaires were cross-checked for additional context and validity of interpretation through semi-structured telephone interviews with a subset of Australian academics who were experienced in engineering education for energy efficiency.

With excellent participation by 27 of the 32 universities teaching higher education (comprising 62 lecturers and 261 students), the survey identified that even though energy efficiency education was highly variable and *ad hoc*, there were a range of preferred options for improvement (Desha et al., 2007; Desha & Hargroves, 2009b). In summary, for more than half of the surveyed courses (55 percent), lecturers reported that their course could include more (in-depth) energy efficiency content, while most respondents (74 percent) thought that the increase in content should be in the specific area of *applying* energy efficiency theory and knowledge. More than half (52 percent) thought their course could include more information about energy efficiency opportunities. The survey also showed a clear preference for resources to be available through open access, online learning modules (90 percent) as opposed to restricted access sources (6 percent) or intensive short courses undertaken in person (13 percent) or remotely (10 percent).

While there was clearly a desire to integrate energy efficiency content, the 2007 Australian survey indicated a substantial shortfall in the inclusion of energy efficiency theory, knowledge, application and assessment in engineering education on the whole. Even mainstream contextual topics such as 'carbon dioxide and other greenhouse gas emissions from energy generation' and 'the link between greenhouse gas emissions and global temperature change' were only covered in detail by up to a third of surveyed courses, and mentioned by less than half. Moreover, student survey results indicated only a low to moderate appreciation of how energy efficiency might be directly related to their future careers. Lecturers and students agreed that there was little if any coverage of topics such as 'product stewardship and responsibility', 'decoupling energy utility profits from kilowatt-hours sold' or 'incremental efficiency versus whole system design'. The survey results indicated that this disconnect – between lecturers recognizing an absence of content, and a lack of action in integrating the content – was likely to be due to the presence of a variety of barriers to implementation. For example, nearly two thirds (58 percent) considered the potential for course content overload to be an issue, while more than half (52 percent)

considered having insufficient time to prepare new materials as a challenge to such curriculum renewal.
This survey contributes to a growing global understanding of the current state of education in this sustainability topic. There is clearly an urgent need to embed energy efficiency knowledge and skills into engineering curriculum, beyond once-off courses, special interest topics in later years, or highly specialized masters programs. These survey findings are also immediately relevant for senior management in engineering departments, Australian professional organizations, and government departments considering future programs and funding allocations, as they provide an indication of the preferred options for increasing energy efficiency education.

3.2 Societal drivers promoting and impeding education for sustainable development

Reports such as the Higher Education Funding Council for England's 2006 report on the 'Barriers and Challenges to Education for Sustainable Development' (Levett-Therivel, 2006) suggest that although actual progress in curriculum renewal has been slow for engineering education, there *is* increasing pressure for curriculum renewal towards engineering education for sustainable development from a range of actors. This includes pressure from the 'top down' (for example from accrediting institutions, professional organizations, advisory boards, education institutions and government) and from the 'bottom up' (for example from faculty members and students themselves). Table 1 provides a brief explanation of the drivers that are promoting such education, synthesizing the literature.

Driver	Factors promoting engineering education for sustainable development
Market/ Business	– *Shifting requirements by potential employers* - increasing requirements for engineers to demonstrate sustainable development capacity. – *Increasing cost of resources and associated taxes/markets* – increasing demand for capacity to reduce water and energy consumption. – *Shifting investment preferences* – increasing attraction to engineers who can reduce energy demand and environmental liabilities. – *Introduction of 'sustainability' rankings* – increasing pressure to improve rankings in indexes (e.g. Dow Jones Sustainability Index). – *Market leadership opportunities* – increasing pressure to achieve/ maintain leadership position and capture early mover advantages. – *Increasing student demand and market potential* - students seeking sustainable development content within their institutions of study.
Information/ Technology	– *Increased scientific understanding* – accumulating scientific knowledge regarding environmental issues, creating pressure for performance improvement in all sectors. – *New technologies* – increasing calls for incorporating a range of new technologies into designs (e.g. renewable energy options). – *New examples of leadership* – emerging examples of leading efforts across sectors will drive competitors. – *Increasing faculty interest in related research and teaching innovation* – increasing incentives offered by governments and organizations. – *Increasing focus in declarations and conference action plans* - creating benchmarks for new kinds of engineering professionals.

Driver	Factors promoting engineering education for sustainable development
Institutional/ Civil Society	– *Shifting accreditation requirements for graduate engineers* - formalising sustainability knowledge and skill requirements. – *Mandatory disclosure and reporting* – increasing disclosure and reporting requirements (e.g. greenhouse gas emissions). – *Increasing professional advocacy* - with leaders stating the pivotal role of engineering in addressing 21st Century challenges. – *Shifting requirements for practising engineers by professional organizations* - where mission statements, code of ethics statements and codes of practice are being updated. – *Increasing commitment and action by highly regarded university peers* - increasingly vocal commitments and alliances.

Table 1. Key factors promoting engineering education for sustainable development

A number of key barriers are also evident in the literature, which appear to be limiting efforts by engineering educators to undertake significant and rapid engineering curriculum renewal, as summarized in Table 2.

Barrier	Factors limiting engineering education for sustainable development
Market/ Business	– *Persistent 'old economy' industry practices,* wherein employers continue to employ graduates to undertake unsustainable practices. – *Uncertainty around future requirements to change* – where varying government messages create considerable uncertainty around impending requirements to change. – *Perceived threat to employability and position,* from taking action ahead of market or sector wide requirements to do so. – *Short-termism in the Higher Education Institution (HEI) sector,* where short-term pressures demand increasing staff to student ratios, and increasing student intake, rather than program innovation. – *A shortage of engineering graduates,* resulting in a 'take what you can get' scenario, to then up-skill internally.
Information/ Technology	– *Growing disconnect between engineering and science,* where engineering professionals may not be 'in-step' in understanding the complexity and interdisciplinary nature of 21st Century challenges. – *Lack of convenient access to emerging and rigorously reviewed information,* where academics may have difficulty getting information and those who have good access may be overwhelmed. – *Lack of access to information in foreign languages,* which may impede the integration of emerging technologies and innovations.
Institutional/ Civil Society	– *Lack of strong requirements for change,* where there is a lack of certainty about current and future legislative requirements and support. – *Lack of academic staff competencies in EESD,* with a relatively low rate of professional development among educators.

Table 2. Key drivers limiting engineering education for sustainable development

Hence, there exist a number of significant societal drivers promoting curriculum renewal within engineering education, which are being tempered by a number of barriers that are limiting the progress. These barriers and others have been strong enough to-date, to prevent a transition towards engineering education for sustainable development in the majority of universities around the world. Many engineering departments are doing little more than including one or two 'sustainability' courses within existing programs, leaving isolated individuals or small teams within departments to undertake *ad hoc* curriculum renewal efforts. In reality, most current engineering degrees are still focused on what could broadly be classified as 'fossil fuel based old industry', involving linear 'heat, beat and treat' processes that don't tend to consider rethinking waste, minimizing inputs, maximizing productivity, capturing synergies or other externalities as part of the process (Benyus, 1997).

3.3 Curriculum drivers promoting and impeding energy efficiency education

Given these observations regarding societal drivers promoting and limiting engineering education for sustainable development, in 2009 the NFEE funded an investigation into identify challenges and opportunities for timely curriculum renewal in energy efficiency education, at the level of the lecturer (Desha & Hargroves, 2009b). Specifically, the project focused on developing and releasing a strategic document to assist the curriculum renewal process for energy efficiency education, drawing upon a behavior change methodology developed by McKenzie-Mohr and Smith (2007). The findings were intended for use by engineering departments, accreditation agencies, professional bodies and government, to identify opportunities for moving forward, and then to strategically plan the transition. The project also provided a significant opportunity to explore options to support lecturers, program co-ordinators and staff to strategically approach, in an informed way, the challenge of increasing the levels of education for energy efficiency as a proxy for other sustainable development topics.

Through a comprehensive literature review followed by a national survey of engineering educators, the researchers short-listed 10 favored options amongst HEIs to integrate emerging energy efficiency content within current engineering programs, as shown below (in order of priority):

1. Including a case study on energy efficiency.
2. Including a guest lecturer to teach a sub-topic.
3. Offering supervised research topics on energy efficiency themes.
4. Offering energy efficiency as a topic in a problem-based learning course.
5. Including assessment that aligns with the energy efficiency theme within the course (e.g. exam questions and assignments)
6. Including tutorials that align with the energy efficiency theme in the course (e.g. presentations/ discussions/ problem solving)
7. Overhauling the course to embed energy efficiency
8. Including one workshop on energy efficiency in the course (i.e. experiments)
9. Including a field trip related to energy efficiency
10. Developing a new course on energy efficiency

Table 2 provides a summary of the identified common barriers to one or more of the shortlisted options, highlighting that putting in place mechanisms to address a particular barrier can have multiple flow-on benefits for addressing other barriers. For example, for key staff who are tasked with integrating new content, setting up an annual allocation of teaching buy-out funds, or having an avenue for temporarily altering staff teaching-research-service workload allocation to engage in rapid curriculum renewal, would help to address the barrier of insufficient time for preparation, which affects 7 of the 10 options. Similarly, an annual small-grants program available for educators to pilot rapid curriculum renewal initiatives would help to address the barrier of prohibitive cost. A 'tiered' approach could be applied, where the first three options, including the use of case studies, guest lecturers and supervised research, may immediately be targeted, with other options then implemented among various programs in the following budget cycles.

Key Issues for Implementation	Shortlisted Options for Curriculum Renewal									
	1.Case Study	2. Guest Lecturer	3. Supervised Research	4. PBL Topic	5. Include Assessment	6. Tutorials	7. Course Overhaul	8. Workshop	9. Field Trip	10. New Course
Common Barriers										
Lack of available data/ information	●	●		●	●	●	●		●	●
Lack of time for preparation	●	●		●	●	●		●		●
An overcrowded curriculum	●		●	●		●			●	●
Prohibitive cost	●		●	●	●	●		●	●	●
Lack of knowledge	●	●	●	●	●		●		●	●
Lack of value attached	●		●			●				
Lack of industry contacts		●	●					●		
Resistance to top-down directive			●				●			
Students' prior learning habits					●				●	
Lecturer apathy		●					●			
Administrative coordination							●	●		●
Common Benefits										
Improved marketability	●	●					●	●		●
Cross-functionality of content	●						●			●
Additional research opportunities		●								●
Networking opportunities for students		●	●					●		
Networking opportunities for lecturers		●	●					●		
Experience in incorporating emerging concepts into curriculum			●				●			
Addressing the time-lag for graduates			●				●			
Improved pedagogy - problem based learning				●	●	●			●	
Improved pedagogy – generic skills				●	●	●			●	
Lecturer professional development		●				●				

Table 1. Identified key barriers and benefits to timely curriculum renewal in energy efficiency education *Source:* (Desha et al., 2009b)

4. Enabling capacity building for energy efficiency

With such considerations in mind, higher education institutions can strategically allocate budget and human resourcing to integrate new content - in this case energy efficiency knowledge and skills - into existing education and training programs. However, the successful transition of engineering education to incorporate such new material is reliant on a number of factors as discussed in the following paragraphs.

4.1 Institutional leadership and support

According to a study by an American campus sustainability assessment project, higher education institutions which are leading in embedding sustainable development knowledge and skills within the curriculum share a number of characteristics: "*First, these 'sustainability leaders' have adopted serious strategies for systematically addressing the sustainability of the institution. They have policies stating their commitment to sustainability goals, and they have specific plans in place that explain how they intend to achieve them. Second, these institutions have provided the resources needed to implement their sustainability plans. They hire staff, form committees, allocate budgets, and show clear administrative support for sustainability initiatives. Third, these sustainability leaders know where they have been, where they are, and where they are headed in terms of sustainability. They measure and track their progress toward sustainability, and regularly meet and update goals and targets*" (The Campus Sustainability Assessment Project, undated).

A 2008 report to the Australian Teaching and Learning Council on addressing the supply and quality of engineering graduates for the new century observed four supporting actions that were common in institutions facilitating significant change, namely: 1) vision; 2) leadership; 3) stakeholder engagement; and 4) resources (King, 2008). Hence, where a period of rapid curriculum renewal is required, it needs to be supported with appropriate resources for the relevant staff members, and undertaken in a realistic timeframe. Staff members need to be encouraged to consider their own strengths and professional development opportunities in contributing to decisions about how their courses embed sustainability knowledge and skills. Existing and proactive efforts by staff in curriculum renewal (i.e. the 'leaders' or 'champions' to date) should be acknowledged, supported and rewarded. A strong collaborative foundation across sub-communities (for example across different disciplines, or different campuses) is also an important mechanism to successfully address surprises or issues as they arise during the curriculum renewal process.

University support could include the provision of funding, marketing and flexibility in rules regarding developing new courses and modifying existing courses. A number of these suggestions involve investing funds, which can be a challenge. However, institutional benefits are clear and in the short term opportunities could be creatively explored for example through industry course sponsorship, the appointment of funded 'sustainability chairs' and professional development bursaries.

4.2 Strategic planning and implementation

For the various curriculum renewal options to be successful, an overarching strategic plan is needed, which maps out timeframes, responsibilities and resource requirements. In the NFEE investigation, a number of key components were identified that might be considered in a strategic plan to rapidly develop graduates who can fill critical energy efficiency knowledge and skills gaps (Desha et al., 2009b):

- Planning from the outset, the best approach for the department given the opportunities and risks with niche degrees versus embedding content throughout programs and offering short courses.
- Building a strong collaborative foundation across campus sub-communities to successfully address surprises or issues as they arise.
- Accessing the growing online library of academically rigorous open-access teaching and learning resources to accelerate course development and renewal;
- Undertaking bridging and outreach opportunities across industry and government, undergraduate and postgraduate programs, and high schools and the community, to recruit students to the renewed programs;
- Making use of national and international collaboration with other academic institutions and non-profit organizations, to jointly deliver courses on energy efficiency topics.
- Integrating such capacity building into campus operations as a two-way collaboration between academics and students.

4.3 Catalysts for accelerating curriculum renewal

To address the existing time lag dilemma evident within engineering education, it is important to set clear timeframes for capacity building processes. Three catalysts that can set such timeframes are briefly discussed here:

- *Program accreditation:* Within regulated disciplines such as engineering, accreditation is a strong driver of change, setting a review period of 3-5 years for universities to continually reflect on and demonstrate how they have addressed existing and emerging accreditation requirements in their programs, in order for their programs to remain endorsed by the accrediting institution. However accreditation is quite a weak driver for engineering education for sustainable development in reality, due to the lack of clear direction on how much or within what timeframe to embed sustainability into engineering curriculum. Furthermore, accreditation agencies and their academic representatives on accreditation committees and boards do not necessarily have adequate understanding of future needs and expectations for curriculum, resulting in a lack of ability to change accreditation requirements. This situation was highlighted more than a decade ago by the Australian Higher Education Council in their report on *Professional Education and Credentialism* (Higher Education Council, 1996), which outlined difficulties facing universities and professional bodies when defining pathways for professional education.
- *Employment:* Both government and industry are significant potential catalysts in their role as current and future employers of undergraduate and postgraduate students, setting clear expectations about changing future employment and training needs. For example, both government and industry could assist professional organizations and the universities themselves (for example through advisory boards) to identify current and future industry demands for graduates with specific knowledge and skill capabilities, and in the demands of undergraduate and postgraduate students themselves. Government and industry could require employees who are undertaking professional development, to include a certain number of hours each year learning about sustainability related technology and innovations.

- *Regulation and policy:* Government can play a role in catalyzing rapid curriculum renewal through providing both penalties and incentives. This could be for example through regulation, requiring industry to accelerate efforts such as energy efficiency assessments. Government could also play a role in influencing professional accreditation requirements to provide the necessary 'calls for action' in priority knowledge and skills areas, to review and revise the coverage and extent of accreditation requirements. Government could change the criteria and selection for research funding, and link a portion of federal funding for higher education institutions to institutional learning and teaching performance with regard to integrating energy efficiency knowledge and skills into curricula.

An example of a government catalyst role can be seen in the example of the Australian federal government's 'Energy Efficiency Opportunities' program, launched in July 2006, which required more than 220 businesses (representing around 45 percent of national energy demand) that use more than 0.5 PJ (approximately 139,000 MWh) of energy per year, to undertake an energy efficiency assessment and report publically on opportunities with a payback period of up to 4 years (DRET, undated). Further to this, Victoria was the first state to require all EPA license holders using more than 0.1 PJ (27,800 MWh) to implement opportunities with a payback period of up to 3 years, through its 'Industry Greenhouse Program' (Victorian Environmental Protection Agency, undated). As a result of implementing these programs, both state and federal government has identified a significant skills shortage in the area of undertaking energy efficiency assessments.

Subsequently the federal government initiated a 'Long Term Training Strategy for the Development of Energy Efficiency Assessment Skills', beginning in 2009 with an extensive survey process across the energy intensive industries, energy service providers, and universities (Council of Australian Governments, 2009). In 2007, the CSIRO (Commonwealth Scientific and Industrial Research Organization) through its 'Energy Transformed Flagship' engaged researchers from The Natural Edge Project to provide capacity building notes for professionals and students looking to up-skill in energy efficiency opportunities, aimed at both undergraduate education and professional development, as discussed below.

5. Capacity building resources

In 2007, the CSIRO funded the development of three education and training modules (30 lectures) in line with its goal for its 'energy transformed' program, '*to facilitate the development and implementation of stationary and transport technologies so as to halve greenhouse gas emissions, double the efficiency of the nation's new energy generation, supply and end use, and to position Australia for a future hydrogen economy*'. It was intended that these modules would provide a base capacity-building training program that would prepare engineers/technicians/facilities managers/architects etc. to address the issues of greenhouse gas emissions and work towards creating sustainable energy solutions throughout the course of their professional life. Within this context the modules would provide an introduction to energy efficiency and low emissions technologies.

The resultant *Energy Transformed* education package (Smith et al., 2007) contains over 600 pages of peer-reviewed content that is freely available online, covering a wide range of issues related to energy for use in undergraduate education, providing industry, business and households with the knowledge they need to realize at least 30 percent energy

efficiency savings as rapidly as possible. The text also provides an updated overview of the latest advances in low carbon technologies, renewable energy and sustainable transport.
The contents of the *Energy Transformed* program are separated into three 'modules':

- *Module A:* Understanding, identifying and implementing energy efficiency opportunities for industrial/commercial users - by technology.
- *Module B:* Understanding, identifying and implementing energy efficiency opportunities for industrial/commercial users - by sector.
- *Module C:* Integrated approaches to energy efficiency and low emissions electricity, transport and distributed energy.

The component chapters and lessons in the package are summarized in Table 4.

Module	Content
A	*Chapter 1: Climate Change Mitigation in Australia's Energy Sector* 1.1: Achieving a 60 percent reduction in greenhouse gas emissions by 2050 1.2: Carbon down, profits up - multiple benefits for Australia 1.3: Integrated approaches to energy efficiency & low carbon technologies 1.4: A whole systems approach to energy efficiency in new & existing systems *Chapter 2: Energy Efficiency Opportunities for Commercial Users* 2.1: The importance & benefits of a front-loaded design process 2.2: Opportunities for energy efficiency in commercial buildings 2.3: Opportunities for improving the efficiency of HVAC systems *Chapter 3: Energy Efficiency Opportunities for Industrial Users* 3.1: Opportunities for improving the efficiency of motor systems 3.2: Opportunities for improving the efficiency of boiler and steam Distribution systems 3.3: Energy efficiency improvements available through co-generation
B	*Chapter 4: Responding to Increasing Demand for Electricity* 4.1: What factors are causing rising peak and base load electricity demand in Australia? 4.2: Demand management approaches to reduce rising 'peak load' electricity demand 4.3: Demand management approaches to reduce rising 'base load' electricity 4.4: Making energy efficiency opportunities a win-win for customers and the utility: decoupling energy utility profits from electricity sales *Chapter 5: Energy Efficiency Opportunities in Large Energy Using Industry Sectors* 5.1: Opportunities for energy efficiency in the aluminum, steel and cement sectors 5.2: Opportunities for energy efficiency in manufacturing industries 5.3: Opportunities for energy efficiency in the it industry and services sector *Chapter 6: Energy Efficiency Opportunities in Light Industry and Commercial Sectors* 6.1: Opportunities for energy efficiency in the tourism and hospitality sectors 6.2: Opportunities for energy efficiency in the food processing and retail sector 6.3: Opportunities for energy efficiency in the fast food industry
C	*Chapter 7: Integrated Approaches to Energy Efficiency and Low Emissions Electricity* 7.1: Opportunities & technologies to produce low emission electricity from fossil fuels 7.2: Can renewable energy supply peak electricity demand? 7.3: Can renewable energy supply base electricity demand? 7.4: Hidden benefits of distributed generation to supply base electricity demand *Chapter 8: Integrated Approaches to Energy Efficiency and Transport* 8.1: Designing a sustainable transport future 8.2: Integrated approaches to energy efficiency & alternative transport fuels - passenger 8.3: Integrated approaches to energy efficiency and alternative transport fuels - trucking *Chapter 9: Integrated Approaches to Energy Efficiency and Distributed Energy*

Module	Content
	9.1: Residential building energy efficiency and renewable energy opportunities: towards a climate-neutral home 9.2: Commercial building energy efficiency and renewable energy opportunities: towards climate-neutral commercial buildings 9.3: Beyond energy efficiency and distributed energy: options to offset emissions

Table 4. Energy Transformed: Sustainable Energy Solutions for Climate Change Mitigation Source: (Smith et al., 2007)

In summary, these modules bring together the knowledge of how countries, specifically Australia, can achieve at least 60 percent cuts to greenhouse gas emissions by 2050, in line with the activities of the CSIRO Energy Transformed Flagship research program which is focused on research that will assist Australia to achieve this target. The materials provide industry, governments, business and households with knowledge to realize at least 30 percent energy efficiency savings in the short term while providing a strong basis for further improvement. It also includes an overview of advances in low carbon technologies, renewable energy and sustainable transport. While the package has an Australian focus, it outlines sustainable energy strategies and provides links to numerous online reports which can assist climate change mitigation efforts globally. It seeks to inform other initiatives that are encouraging the reduction of greenhouse gas emissions, for example through behavior change, sustainable consumption, and changes to economic incentives and policy.

The online format of this education and training program has been designed using the results of the 2007 and 2009 NFEE funded research, including the following considerations:

- Extensive peer review was sought during the writing process (see acknowledgements) to assist with creating awareness of the materials.
- The length of content (i.e. number of pages) for each lecture is intended to make it easy for rapid uptake, using a 20-30 page highly structured and straightforward format that suits most learning environments.
- Each lecture provides links to numerous online reports that outline sustainable energy strategies and which assist climate change mitigation efforts in Australia and globally.
- The presentation of the content has been designed for flexibility, to cater for a range of learning processes, from self-paced modular learning through to PowerPoint presentations, tutorial discussions and problem-based learning.

Each 'lecture' begins with an 'Educational Aim' which provides an overview of the module. This is followed by a section called, 'Essential Reading', wherein key references used in the module that is readily accessible (i.e. with regard to language and layout) are listed and hyperlinked where practical. The lecture then proceeds with around ten 'Learning Points' that are around 3-4 sentences each, which step through the core knowledge. Each learning point has been worded so that key words can be easily extracted for PowerPoint slides or handouts. Following the learning points, the lecture includes several pages of 'Background Information', which provide both contextual information and deeper insights into the knowledge area. This is also intended to provide a straightforward and short briefing to lecturers/ trainers who may not have prior knowledge of the specific content. Following the learning points, a list of 'Optional Reading' is provided as an additional resource for assignments or further research, and a set of 'Key Words for Searching Online' are listed to assist with beginning an internet exploration.

6. Conclusion

This chapter has discussed the need for urgent capacity building in the engineering profession in the area of energy efficiency, focusing on higher education institutions. We have considered the complexity of the issue within the higher education sector, where the problem is two-fold: energy efficiency knowledge and skills are not yet being taught; and the process for curriculum renewal is generally slow and *ad hoc*. Moreover, there are a number of organisational and curriculum influences that are working to both promote and impede capacity building in energy efficiency, requiring a strategic and systematic approach to ensure that the engineering profession is being up-skilled as quickly as possible. This includes leadership and support at an institutional level, strategic planning and implementation of curriculum renewal initiatives, and within clear timeframes. Three examples of energy efficiency capacity building initiatives in Australia have been highlighted; namely two research initiatives undertaken through the National Framework for Energy Efficiency (NFEE), and modular content developed through the CSIRO's energy transformed program.

In conclusion, the research undertaken to date provides a clear understanding of the state of engineering education for energy efficiency in Australia. Furthermore, the *Energy Transformed* education package funded by the CSIRO provides a significant tool for engineering educators to access, to provide immediate and robust capacity building, from undergraduate through to postgraduate education.

7. Acknowledgements

The 2007 survey, 2009 investigation and the *Energy Transformed* education package were undertaken by The Natural Edge Project using funds provided by CSIRO and the National Framework for Energy Efficiency. Non-staff related on-costs and administrative support was also provided by the Centre for Environment and Systems Research and the Urban Research Program at Griffith University, and the Fenner School of Environment and Society and Engineering Department at the Australian National University.

Principal reviewers for the three research initiatives included: Adjunct Professor Alan Pears – RMIT, Geoff Andrews – Director, Genesis Now Pty Ltd, Dr Mike Dennis – ANU, Engineering Department, Victoria Hart – Basset Engineering Consultants, Molly Olsen and Phillip Toyne - EcoFutures Pty Ltd, Glenn Platt – CSIRO, Energy Transformed Flagship, and Francis Barram – Bond University. The following persons provided peer review for specific lectures in the *Energy Transformed* program; Dr Barry Newell – Australian national University, Dr Chris Dunstan - Clean Energy Council, D van den Dool - Manager, Jamieson Foley Traffic & Transport Pty Ltd, Daniel Veryard - Sustainable Transport Expert, Dr David Lindley – Academic Principal, ACS Education, Frank Hubbard – International Hotels Group, Gavin Gilchrist – Director, BigSwitch Projects, Ian Dunlop - President, Australian Association for the Study of Peak Oil, Dr James McGregor – CSIRO, Energy Transformed Flagship, Jill Grant – Department of Industry Training and Resources, Commonwealth Government, Leonardo Ribon – RMIT Global Sustainability, Professor Mark Diesendorf – University of New South Wales, Melinda Watt - CRC for Sustainable Tourism, Dr Paul Compston - ANU AutoCRC, Dr Dominique Hes - University of Melbourne, Penny Prasad - Project Officer, UNEP Working Group for Cleaner Production, University of Queensland,

Rob Gell – President, Greening Australia, Dr Tom Worthington - Director of the Professional Development Board, Australian Computer Society.

8. References

Australian Industry Group (2007). *Environmental Sustainability and Industry, Road to a Sustainable Future*, AIG, Australia.

Australian Federal Department of Resources, Energy and Tourism (DRET) (undated) 'Energy Efficiency Opportunities', http://www.ret.gov.au/energy/efficiency/eeo/pages/default.aspx, accessed 12 May 2010.

Benyus, J. (1997). *Biomimicry: Innovation Inspired by Nature*, HarperCollins Publishers Inc., New York.

Brown, L. (2007). *Plan B 3.0: Mobilising to Save Civilisation*, W.W. Norton & Company, New York, 117 and 208.

Council of Australian Governments (2009). *National Strategy on Energy Efficiency*, July 2009, Commonwealth of Australia.

Desha, C. & Hargroves, K. (2009a). Re-engineering Higher Education for Energy Efficiency Solutions. *ECOS*, CSIRO, Vol. 151, (Oct-Nov 2009) 16-17, 0311-4546.

Desha, C. & Hargroves, K. (2009b). Surveying the State of Higher Education in Energy Efficiency in Australian Engineering Curriculum. *Journal of Cleaner Production*, Elsevier, Vol. 18, Issue. 7, 652-658, 0959-6526.

Desha, C. ; Hargroves, K. & Smith, M. (2009). Addressing the time lag dilemma in curriculum renewal towards engineering education for sustainable development, *International Journal of Sustainability in Higher Education*, Vol. 10, No. 2, 184-199, Elsevier.

Desha, C. ; Hargroves, K. & Stephens, R. (2007). *Energy Transformed: Australian University Survey Summary of Questionnaire Results*. The Natural Edge Project (TNEP), Australia.

Desha, C. ; Hargroves, K. & Reeve, A. (2009). *An Investigation into the Options for Increasing the Extent of Energy Efficiency Knowledge and Skills in Engineering Education*, Report to the National Framework for Energy Efficiency, The Natural Edge Project (TNEP), Australia.

Energy Futures Forum (2006). *The Heat Is On: The Future of Energy in Australia*, CSIRO

Garnaut, R. (2008) *The Garnaut Climate Change Review*, Report to the Australian Federal Government, Canberra.

Hatfield-Dodds, S. ; Turner ,G. ; Schandl, H. & Doss, T. (2008). *Growing the green collar economy: skills and labour challenges in reducing our greenhouse emissions and national environmental footprint*, Report to the Dusseldorp Skills Forum, CSIRO Sustainable Ecosystems, Canberra.

Higher Education Council (1996). *Professional Education and Credentialism*, National Board of Employment, Education and Training (NBEET), Canberra, p4.

IPCC (2007) *Climate Change 2007: Impacts, Adaptation & Vulnerability, Contribution of Working Group II to the Fourth Assessment Report of the Intergovernmental Panel on Climate Change*, Cambridge University Press, Cambridge.

King, R. (2008). *Addressing the Supply and Quality of Engineering Graduates for the New Century*, The Carrick Institute for Learning and Teaching in Higher Education Ltd, Sydney, p30.

Levett-Therivel (2006). *Sustainability carrots and sticks: the benefits and risks of sustainable development for HEIs*, Report to the Higher Education Funding Council of England (HEFCE).

McKenzie-Mohr, D. & Smith, W. (2007). *Fostering Sustainable Behaviour: An Introduction to Community-Based Social Marketing*, 3rd Ed, New Society Press, 0-86571-406-1, Gabriola Island, BC.

Postel, S. (1999). *Pillar of Sand*, W.W. Norton & Company, New York, 13-21.

Price Waterhouse Coopers (2008). *Carbon countdown: a survey of executive opinion on climate change in the countdown to a carbon economy*, PWC.

Proudfoot Consulting (2007). *Meeting the Corporate Energy Challenge: Are Companies Walking the Talk on Energy Efficiency?*, Proudfoot Consulting.

Smith, M. ; Hargroves, K. ; Stasinopoulos, P. ; Stephens, R. ; Desha, C. & Hargroves, S. (2007). *Energy Transformed: Sustainable Energy Solutions for Climate Change Mitigation*, The Natural Edge Project (TNEP), Australia.

Stern, N. (2006) *The Stern Review: The Economics of Climate Change*, Cambridge University Press, Cambridge.

The Campus Sustainability Assessment Project (undated) 'Introduction: The Sustainability Imperative and Higher Education: The Challenge of Sustainability', http://csap.envs.wmich.edu/pages/intro_imperative.html, accessed 13 May 2010.

The Institution of Engineers Australia (undated) 'Energy Efficiency: The Importance of Energy Efficiency in Moving toward Sustainability', http://www.engineersaustralia.org.au, accessed 12 May 2010.

United Nations (2002) *Report of the World Summit on Sustainable Development*, United Nations, Johannesburg, South Africa, 26 August – 4 September 2002.

United Nations Environment Programme (2008) *Green Jobs: Towards Decent Work in a Sustainable, Low-Carbon World*, United Nations.

US Geological Survey (2007), *Mineral Commodity Summaries 2007*, cited in Brown, L. (2007) *Plan B 3.0: Mobilising to Save Civilisation*, W.W. Norton and Company, New York, 43.

Victorian Environmental Protection Agency (undated) 'Environment and Resource Efficiency Plans – EREP', http://www.epa.vic.gov.au/bus/erep/, accessed 12 May 2010.

World Resources Institute (2005) *World Resources 2005: The Wealth of the Poor-Managing Ecosystems to Fight Poverty*, WRI, in collaboration with the UNDP, UNEP and the World Bank, Washington, DC.

8

The energy efficiency of onboard hydrogen storage

Jens Oluf Jensen, Qingfeng Li and Niels J. Bjerrum
Technical University of Denmark
Denmark

1. Introduction

Hydrogen is often suggested as a versatile energy carrier in future energy systems. Hydrogen can be extracted from water by electrical energy through electrolysis and later when the energy is needed hydrogen can recombine with oxygen from the air and release the same amount of energy. The end product is water and the cycle is closed. Hydrogen as an energy carrier is typically associated with renewable energy technologies because it provides a way to store energy. The need for energy storage is tremendous if wind turbines, wave energy devices or photovoltaics are to be implemented on a large scale. This is because of the fluctuating nature of the electricity production by these means and moreover, because the energy might be meant for application in the transport sector. Batteries store electrical energy efficiently, but they are not economic for large scale storage and for transportation they are only practical in smaller vehicles with a limited driving range and certainly not in trucks, ships or airplanes. The alternative is to store energy as hydrogen and hydrogen is an ideal fuel in many aspects. It is easily combusted in an engine or converted back to electricity in a fuel cell. It is not poisonous and the raw material for its production (water) is practically unlimited.

Hydrogen is often said to have the highest energy content per unit mass, but since it is a low density gas at ambient conditions it needs a storage tank that adds so much to the weight and volume that the whole system ends up being both heavier and bulkier than a gasoline tank with the same energy content. Therefore, hydrogen storage is a key issue, and in particular, onboard hydrogen storage in vehicles. As a matter of fact, the production of hydrogen from renewable sources only makes sense if hydrogen is stored for later use or for use elsewhere. Otherwise, one might as well use the extracted electricity directly (one exception could be the use of bio fuels in a fuel cell through a stage where hydrogen is liberated by reforming for immediate use, but this is not really within the idea of hydrogen as an energy carrier). Two recent monographs each provide a detailed introduction to all the aspects of hydrogen energy with several chapters dealing with storage techniques (Leon, 2008), (Züttel et al., 2008).

Many different techniques have been developed to solve this fundamental problem, and any one of them has its own energy balance to consider. Storage of hydrogen can be quite energy consuming and so can the subsequent liberation of hydrogen. In some cases both processes are energy intensive. The literature on hydrogen storage often focuses on the storage density, and the question of round trip energy efficiency of the storage process may then be forgotten. In small systems, such energy losses might, although significant, be of less importance, but for vehicular applications, they cannot be neglected. After all, improved efficiency is one of the arguments when future fuel cell vehicles are compared with conventional ones. This work will review the most common hydrogen storage techniques with the focus on energy efficiency for charging and discharging the system, i.e. the round trip efficiency. It is an elaborated version of a previous study (Jensen et al., 2007).

2. Overview of storage techniques

Hydrogen is a volatile gas at ambient conditions, and the storage challenge is to fight the kinetic energy of the hydrogen molecules. Basically there are three ways to go. (1) The gas can be confined at high pressure by external physical forces. (2) The energy of the molecules can be withdrawn by cooling and ultimately the gas condenses into a liquid. (3) The molecules can be bound to a surface or inside a solid material. This way hydrogen is more or less immobilized and like in the case of liquid hydrogen, most of its kinetic energy is removed. The three fundamental storage techniques are visualised in the corners of the triangle in figure 1. Between the corners combined techniques that utilize more than one of the principles are plotted.

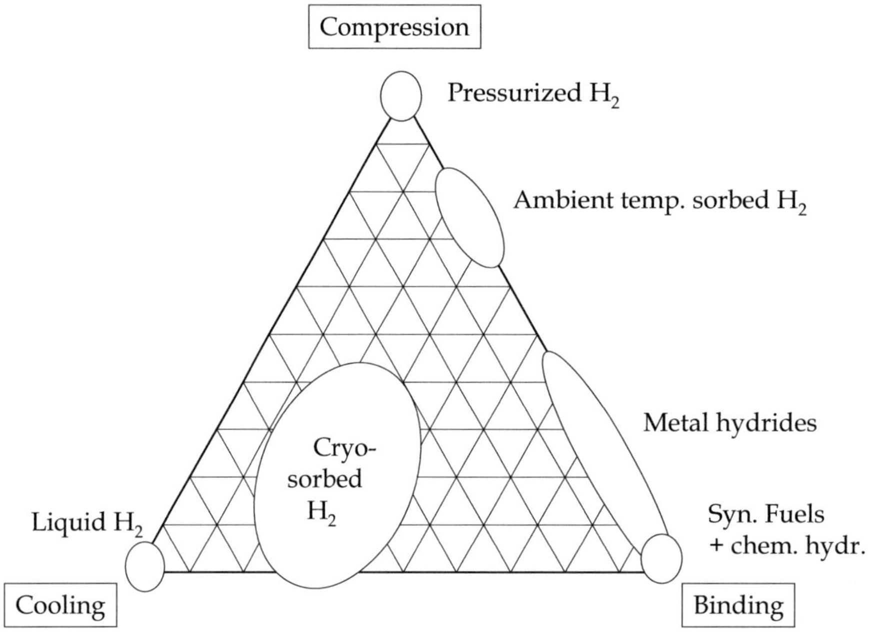

Fig. 1. The different storage techniques arranged qualitatively after degree of cooling, binding and pressurization.

Compressed hydrogen is kept in a dense state by external physical forces only. This is what happens in a pressure vessel. It takes mechanical energy to compress the gas, but the release is free of charge. Liquid hydrogen is kept together by weak chemical forces (van der Waals)

at very low temperature but at ambient pressure. Heat must be supplied to release hydrogen through boiling, but due to the low boiling point of 20 K, the heat can in principle be taken from the surroundings or any waste heat. Liquefaction of hydrogen by pressurization alone is not possible since the critical point is as low as 33 K (and 13 bar).
Hydrogen can bind to matter in many ways. It can be via adsorption on a large surface with some affinity for hydrogen molecules. In order to obtain a reasonable storage capacity this is always done in combination with either cooling (to reduce the energy of the hydrogen molecules), pressurization or both. The binding forces are the weak van der Waals forces like in liquid hydrogen, but the interaction is stronger due to the substrate. Release is comparable to a combination of compressed and liquid hydrogen. Absorption of hydrogen takes place in specialized solid materials into which hydrogen can diffuse and bind by metallic, ionic or covalent bonds. These forces are much stronger than the van der Waals forces and consequently, it takes more energy to release hydrogen afterwards. Examples are interstitial metal hydrides and complex hydrides. Finally it is possible to store hydrogen by making synthetic fuels like hydrocarbons, alcohols and ammonia. In this case the bonds are mostly covalent and require a significant amount of energy for hydrogen release. Moreover, in many cases, addition of water is needed too like for steam reforming. Synthetic fuels cannot be recharged onboard. Instead they are manufactured through chemical synthesis in a plant.
Another way to arrange the storage techniques is shown in figure 2, where they are ordered in a line ranging from pure physical storage to a gradually more chemical technique. A tendency that goes with this is that the more chemical the technique, the less easily available is the hydrogen. This less easy availability of hydrogen is seen as higher energy demands for hydrogen release and/or higher release temperatures.

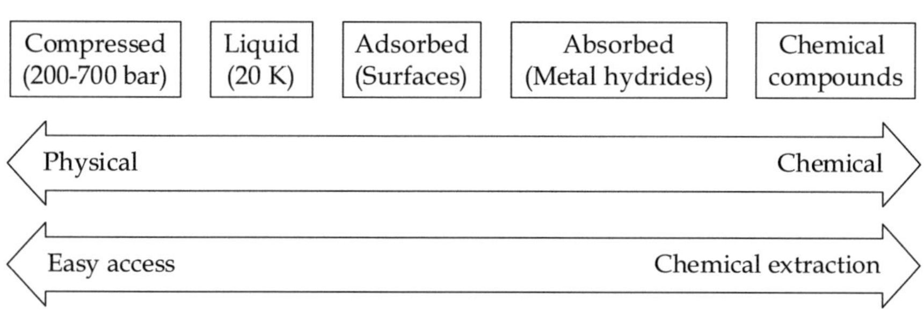

Fig. 2. The sequence of hydrogen storage techniques from physical to increasingly chemical.

3. The approach

The different storage techniques are in the following treated in the same order as in figure 2 from left to right. Although hydrogen storage does in principle not depend on the application, onboard storage, e.g. on a vehicle, is assumed since here we have the most demanding situation that may justify sophisticated and possibly expensive storage techniques. The aim of the study is first of all to compare the minimum energies required for storing hydrogen and releasing hydrogen. When energy is needed for the release, typically heat, it can in some cases be supplied by otherwise wasted heat from an engine or a fuel cell, but it depends on the temperature of that heat whether it is possible. Alternatively, the heat for release can be supplied by part of the hydrogen via a burner. In the latter case the available hydrogen for the main purpose (e.g.

propulsion) will be reduced comparatively and the effective storage capacity is thus lower than predicted from the amount of hydrogen stored.
A true comparison would involve a detailed analysis of whole systems. Such analyses are truly relevant but also complicated with numerous assumptions on which the outcome will strongly depend. Instead, transparency is aimed at with the hope that the conclusions are less questionable, although they do not tell the whole story. Throughout, the lower heating value (LHV) of the fuel is used instead of the higher heating value (HHV). This is because in several of the systems, heat for hydrogen liberation must be supplied at temperatures above 100ºC likely by combustion of hydrogen. It is also assumed that hydrogen or a hydrogen mixture is released at no less than ambient pressure. The LHV used is 242.8 kJ/mol H_2.

4. Compressed hydrogen

Despite many attempts to develop advanced techniques for compact, practical and safe hydrogen storage, pressurization is still the dominating technique. This is a fact for onboard hydrogen as well as for hydrogen storage in general. The standard pressure for steel cylinders is 200 bar, but high pressure fibre composite tanks rated for up to 7-800 bar have been developed. The gravimetric storage capacity ranges from 1-2 wt.% for 200 bar steel tanks to 5-10 wt.% for high pressure fibre tanks. Fibre tanks are more expensive than steel tanks.

4.1 Energy for storage

The theoretical minimum work needed for gas compression can be calculated based on integration of the infinitesimal pressure-volume work, dw

$$dw = Vdp \tag{1}$$

where V is the tank volume and p the pressure. Assuming ideal gas behaviour integration of (1) from p_0 to p_1 results in the expression of the work, W, of ideal isothermal compression

$$W = p_0 V \ln\left(\frac{p_1}{p_0}\right) \tag{2}$$

where p_0 and p_1 are initial and final pressures. At hydrogen pressures over 100 bar, deviations from ideality become significant in this connection, and the dimensionless compression factor, Z, shall compensate for the non-ideality. The real gas equation is then

$$pV = ZnRT \tag{3}$$

Z depends on both pressure and temperature and is tabulated elsewhere (Perry et al., 1984). At 300 K and pressures up to 1000 bar, the compression factor is modelled well as

$$Z = 1 + k_{z,300}\left(\frac{p}{p^{\circ}}\right) \tag{4}$$

where $k_{z,300} = 0.000631$. Integration including (3) and (4) gives

$$W = p_0 V_0 \left[k_{z,300}(p_1 - p_0) + \ln\left(\frac{p_1}{p_0}\right) \right] \tag{5}$$

However, the compression is never isothermal, as heat is formed during the process. If the compression is very slow, most heat will dissipate to the surroundings, but in practical high pressure systems, a significant amount of heat is formed. The other extreme is adiabatic compression in which all heat produced is kept in the gas by ideal insulation. The work of adiabatic compression is

$$W = \frac{\gamma}{\gamma - 1} p_0 V \left[\left(\frac{p_1}{p_0}\right)^{\frac{\gamma-1}{\gamma}} - 1 \right] \tag{6}$$

where γ is the ratio of specific heats (C_p/C_v). $\gamma = 1.41$ for hydrogen. The work of adiabatic compression to a fixed final density is much larger than the work of isothermal compression because the heat accumulated creates a higher pressure for the compressor to work against. Both isothermal and adiabatic compression is plotted in figure 3 as a function of the final pressure. Isothermal compression is the absolute minimum theoretically possible, and in reality, due to the discussed heat effect compression is performed in multiple stages with inter-cooling of the gas. Consequently, the work of compression lies somewhere between the two curves. The efficiency of a compressor system varies a lot and the curve in figure 3 is assuming a satisfactory compressor technology (Bossel et al., 2003).

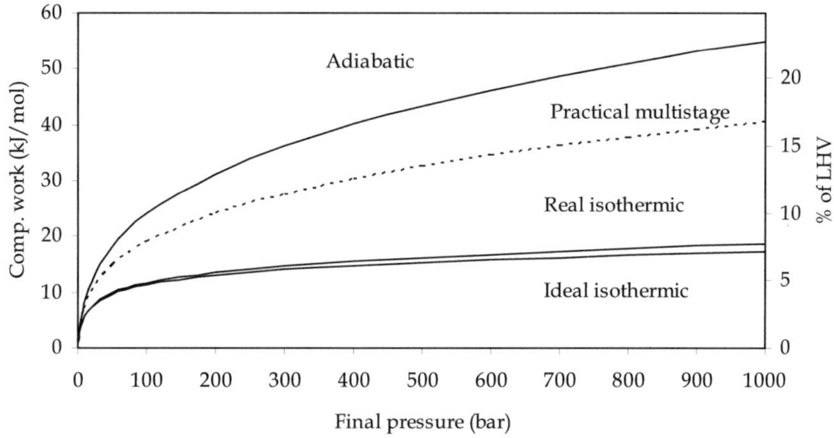

Fig. 3. The energy required to compress hydrogen from 1 bar to the final pressure specified on the primary axis. Re-plotted from (Jensen et al., 2007)

4.2 Energy for release

One strong advantage of compressed hydrogen is that it is easily available at a pressure high enough for fast transport through tubes. Even though the pressure vessel will cool during release, the pressure will in most cases still be way above ambient pressure. Therefore, no energy is needed for the release. In principle, part of the compression energy can even be reclaimed via an expander, but as it adds to complexity and cost it can be argued whether or not it is feasible.

4.3 Discussion

The work of compression in real systems is estimated by Bossel et al (Bossel et al., 2003) and Weindorf et al (Weindorf et al.,2003). According to these studies, compression to 800 bar is possible using 18 % (Bossel) or 13 % (Weindorf) of LHV. The estimated curve for a real system added to figure 3 is between these values. Compression to a final pressure of 800 bar then costs 15.5 % of LHV.

One way to minimize the work of compression is to produce hydrogen by high-pressure electrolysis. The extra voltage (corresponding to extra energy) for reducing hydrogen at high pressure is close to the theoretical value because the reaction kinetics is very fast.

It is evident from equation (2) that the minimum ideal work of compression of one mole hydrogen from 100 to 1000 bar is the same as from 1 to 10 bar. This means that there is a significant benefit even if the electrolyser is operated at just 10-50 bar. Industrial electrolysers working at 32 bar are commercially available today from Statoil (former Norsk Hydro) and IHT.

There may be additional sources for spending energy during filling of the tank. Filling stations often store the gas to higher pressure than used onboard in order to facilitate fast transfer. This "over compression" will naturally lead to some losses. Moreover, cooling may be applied during the transfer to the vehicle and that costs energy too. These effects are not considered in the above calculations.

5. Liquefied hydrogen

Liquid hydrogen has the advantages that it is quite dense and that fuelling is fast and in principle as easy as for gasoline. The main drawbacks are that liquefaction is very energy intensive and that hydrogen continuously evaporates due to influx of heat. The latter can be reduced to a few percent per day or less by advanced thermal insulation, but it will always have to be dealt with. Liquid hydrogen tanks are high cost items and at present liquid hydrogen are only available in selected countries.

5.1 Energy for storage

Some gasses like propane and butane can be liquefied at room temperature by compression to moderate pressures. Unfortunately this is not the case for hydrogen (as well as for oxygen and nitrogen). The reason is that the critical point is situated at a temperature lower than ambient temperature. The critical point of hydrogen is at 33 K and 13 bar. At this point in the phase diagram the gas-liquid equilibrium line ends and above the critical temperature the substance will never liquefy, but will act as a compressed gas even at extreme pressure (eventually hydrogen will form a solid, but at room temperature it requires several thousand bar). This means that for the liquefaction of hydrogen cooling at minimum to 33 K is mandatory.

A simple theoretical pathway for liquid hydrogen is to cool it from room temperature (298 K) to the boiling point at 20 K and then condense it. The average heat capacity in the interval is 28.48 J/mol K, and the heat of vaporization at 20 K is 892 J/mol H_2 (Values by Air Liquide). Based on this, the minimum energy required is 8.81 kJ/mol H_2. To this value 1.06 kJ/mol H_2 should be added for ortho-para conversion of the hydrogen (see below) and the total theoretical enthalpy change is then 9.87 kJ/mol H_2 or 4 % of LHV. In reality the process is quite complicated and much more energy intensive.

Pressurized nitrogen can be liquefied by Joule-Thomson expansion through a valve. The case of hydrogen is more complicated since the Joule-Thomson coefficient [$\mu_{JT} = (\partial T/\partial p)_H$] for hydrogen is negative at room temperature. This means that the gas would heat up instead of cooling. The final complication is that molecular hydrogen exists in two forms called *ortho* and *para*-hydrogen depending on the nuclear spin being parallel (ortho-H_2) or anti-parallel (para-H_2). At room temperature hydrogen contains 75% ortho and 25% para-H_2 and at 20 K the stable form is para-H_2. Hydrogen converts slowly between the two forms and this process is exothermic in the direction ortho-para (-1.06 kJ/mol H_2 at 20 K). Over time the equilibrium composition will be reached, but this takes a time orders of magnitude longer than the liquefaction process. Consequently, heat will be produced in the liquid hydrogen if this is not dealt with in the liquefaction process. The ortho-para conversion can be accelerated by a catalyst, but the heat produced adds to the amount of heat that has to be removed in the liquefaction process.

Practical hydrogen liquefaction plants provide the initial cooling of moderately compressed hydrogen by a conventional cooling system. Further cooling is done by liquid nitrogen followed by a mechanical expander and finally a Joule-Thomson valve. At low temperature the Joule-Thomson coefficient becomes positive and hydrogen cools and liquefies. Along this pathway ortho-para conversion catalysts are inserted at selected locations.

Bossel et al. (Bossel et al., 2003) refer to a detailed analysis concluding that the theoretical energy demand is 14.2 MJ/kg H_2. This equals 28.4 kJ/mol H_2 or 11.7 % of LHV.

In addition, hydrogen needs to be purified to high purity prior to liquefaction because all gaseous impurities will solidify at the low temperature and possibly block the Joule-Thomson valve.

5.2 Energy for release

Because the temperature is very low compared to the surroundings, all heat for hydrogen evaporation should be available. Nevertheless, in practical systems, a built-in electrical heater is often used because heat transfer fluids freeze if passed through heat exchange tubes in the tank. In this study it is assumed that heat from the surroundings is used. However, if electrical heating is used the latent heat of 892 J/mol H_2 should be sufficient to liberate hydrogen gas. Further heating can take place in the external tubing.

5.3 Discussion

The practical energy demand for liquefaction is significantly larger and depends on the size of the plant. Today, the energy demand in a modern plant is on the order of 40-45 % of LHV, but according to Bossel (Bossel et al., 2003), 25 % and to Weindorf (Weindorf et al., 2003), 21% of LHV should be possible in very large liquefaction plants.

6. Adsorbed hydrogen

Like any gas, hydrogen can absorb on surfaces. The molecules are held by the weak van der Waals forces which are much smaller than those of real chemical bonds. Equilibrium is established between adsorbed and free molecules and the surface coverage increases with the gas pressure and with decreasing temperature. Materials like active carbon, carbon nano-tubes, zeolites and metal-organic frameworks have been studied for sorption capacities. The binding energy is 1-10 kJ/mol H_2. It is the general experience that at room temperature and pressures in the range 50-100 bar only up to 1 wt.% hydrogen storage is possible. At liquid nitrogen temperature (77 K) 4-6 wt.% has been reported by different groups. These values refer to the high surface area material only and do not take into account a pressure tank or the insulation in case of cryogenic sorption.

The heat for adsorption is limited compared to the hydrides (treated below). This is an advantage when filling the tank since only 1-10 kJ/mol H_2 is released (0.4-4 % of LHV). However, in the cryogenic case this heat must be removed at 77 K and that requires more energy than for cooling at room temperature or at elevated temperature.

A detailed calculation of the energies for storage and release is very complex. The sorption energy and the storage pressure can both vary by an order of magnitude and if liquid nitrogen is applied it requires an additional energy contribution that too depends on the conditions. When, on top of that, hydrogen storage by adsorption has not yet shown advantages over the other techniques discussed here, the calculation was not attempted.

7. Reversible metal hydrides

The term "reversible hydride" refers to both interstitial and complex hydrides as long as they can be charged as well as discharged by direct solid/gas reactions (or liquid/gas). "Reversible" should not be understood in a thermodynamic sense in this context, it only means "capable of reversing".

Hydrogen stored in interstitial metal hydrides is bound into interstitial positions in a host metal alloy in a more or less metallic way. This bond is stronger than the van der Waals forces mentioned before and a significant amount of heat is required to release hydrogen.

In the complex hydrides or other real chemical systems chemical bonds ranging from ionic to covalent are formed between hydrogen and the carrier atoms. The hydrogen release reactions in these cases typically require a significant energy input and also elevated temperatures to overcome the activation energy.

7.1 Interstitial hydrides

Interstitial hydrides are the most studied metal hydride systems for hydrogen storage. Examples are plentiful such as $LaNi_5H_6$, $TiFeH_{\sim 2}$, and $LaNi_5$-based alloys for nickel metal hydride batteries. They are considered very safe and easy to operate, and their main drawback apart from the price in some cases is the fact that the hydrogen storage capacity (with a few exceptions) is below 2 wt. %. One convenient characteristic is that the alloys can be tailored to a moderate equilibrium pressure of a few bars at ambient temperature. The heat of desorption is then around 30 kJ/mol H_2 or 12 % of LHV. During charging, this heat is liberated. In small canisters, the heat can be exchanged with the surroundings, but in larger systems like in a vehicle, active cooling by water

is necessary. The energy balance of such a cooling system depends highly on the charging rate aimed at. Consequently, only the sorption energy is considered.
When hydrogen is liberated, the hydride cools and the plateau pressure must still be above ambient pressure to avoid subsequent compression of the released hydrogen. This implies that the plateau pressure will be correspondingly higher when the hydride is heating up during charging and the charging pressure must match that. A 20-50 bar charging pressure can be suggested. Based on the discussion above, compression to 20 bar is set to 4-5 % of LHV (or 3 % with isothermal compression).
The amount of heat for desorption is the same as for absorption. It can be taken from the excess heat of the fuel cell or combustion engine provided that the temperature is high enough. The interstitial hydride can be designed for that.

7.2 Other reversible hydrides

Other reversible hydrides obey the same thermodynamic laws but possibly with other pressure-temperature characteristics. They are not as easily tailored, and the reaction enthalpy is generally more or less fixed.
Examples of other reversible metal hydrides are MgH_2, Mg_2NiH_4 and $NaAlH_4$. The two magnesium-based hosts are both characterized by one flat plateau, while $NaAlH_4$ desorbs hydrogen in two steps with different stabilities (Bogdanovic, 2000). The first step is:

$$NaAlH_4 \leftrightarrow 1/3Na_3AlH_6 + 2/3Al + H_2 \quad (7)$$

and the second step is

$$Na_3AlH_6 \leftrightarrow 3NaH + Al + 3/2H_2 \quad (8)$$

The key desorption properties of the mentioned hydrides are listed in table 1. The column "Temperature for 1 bar" is based on thermodynamics. For kinetic reasons, $NaAlH_4$ needs temperatures of around 150°C even when Ti-doped. This means that the charging hydrogen pressure must be on the order of 100 bar which, assuming isothermal compression, takes 4.8 % of LHV or practically 7-8 % of LHV in reality. The other systems can be charged at low pressures like the interstitial hydrides.

Hydride	Rev. capacity	Heat of desorption	Temperature for 1 bar
Interstitial MH	1-2 wt.%	~ 30 kJ/mol H_2 (~12.4 % of LHV)	Near room temperature
MgH_2	7.6 wt.%	74.5 kJ/mol H_2 (30.8 % of LHV)	300°C
Mg_2NiH_4	3.6 wt.%	64.5 kJ/mol H_2 (26.7 % of LHV)	255°C
$NaAlH_4$ (one step)	3.7 wt.%	37 kJ/mol H_2 (15.3 % of LHV)	35°C
Na_3AlH_6	1.9 wt.%	47 kJ/mol H_2 (19.4 % of LHV)	110°C
$NaAlH_4$ (two steps)	5.6 wt.%	40 kJ/mol H_2 (16.5 % of LHV)	110°C

Table 1. Selected metal hydrides and their hydrogen storage properties.

7.3 Energy for storage

The energy for storage is basically the energy for pressurizing hydrogen to the charging pressure, i.e. 10-50 bar for interstitial metal hydrides and 100 bar for $NaAlH_4$ as explained above. Energy for active cooling is not estimated, but will be relevant, especially in case of fast charging.

7.4 Energy for release

The energy for hydrogen release is the heat of desorption as listed in Table 1.

7.5 Discussion

The significant amount of heat liberated during charging will practically need to be removed actively in a vehicle size tank for an acceptable filling rate. There is a separate energy balance for this, but since it can be done in many different ways and at different rates it has been omitted in this context.

8. Irreversible hydrides

8.1 $NaBH_4$

$NaBH_4$ does not easily liberate hydrogen like the hydrides discussed so far, but it reacts with water over a catalyst. $NaBH_4$ is stored in an alkaline aqueous solution in which it is stable. When passed over a catalyst the following reaction takes place

$$NaBH_4 + 2H_2O \rightarrow NaBO_2 + 4H_2 + \text{Heat} \tag{9}$$

The reaction is exothermic with the enthalpy -212 kJ/mol $NaBH_4$ or -53 kJ/mol H_2 (22% of LHV). The hydrogen storage capacity is 21.2 wt.% disregarding the water, but the practical capacity is much lower due to the water. Besides the role as a reactive solvent the water also acts as a heat sink for the heat liberated during the process. The system is commercialized by Millennium Cell®, and several demo cars have been fitted with such a system. The concept can also be used directly in alkaline fuel cells with the catalyst being the anode catalyst (Li et al., 2003).

8.2 Energy for storage

Being irreversible, $NaBH_4$ must be regenerated through other chemical pathways. As a minimum, the 212 kJ/mol must be supplied during that process, but the real number is significantly larger and depends on how regeneration is done.

8.3 Energy for release

No energy is needed for hydrogen liberation apart for pumping the liquid or for active heat management.

8.4 Discussion

The round trip energy efficiency of this hydrogen storage system will most likely exclude it from any application in which energy efficiency matters.

9. Methanol and ammonia

In this group, the hydrogen evolution reactions are characterized by equlibria with both reactants and products in the gas phase. There is no such thing as a desorption temperature at which the hydrogen pressure is 1 bar. The minimum hydrogen release temperature is therefore chosen as the temperature at which kinetics are reasonably fast and the equilibrium is strongly in favour of hydrogen formation.
The liberated hydrogen is in these cases mixed with either carbon dioxide or nitrogen. This fact affects the way a fuel cell is fuelled. As the fuel part of the mixture is consumed, the inert gas fraction increases, and this dilution effect can lead to local starvation of the electrode and poor performance (and lead to electrode degradation). To overcome this problem, fuel is fed in excess of at least 20 % (This problem can to some extent apply to any fuel cell operating below the boiling point of water because of water vapour accumulation followed by condensation. However, it can be solved by eventual purging without large losses). The over-stoichiometry is labelled λ. $\lambda = 1$ means strictly stoichiometric and $\lambda = 1.2$ means 20% excess. The 20 % excess fuel is normally combusted in a burner, and the resulting heat can then be used for fuel processing.

9.1 Methanol (CH_3OH)

In order to extract hydrogen from methanol it can be steam reformed according to

$$CH_3OH + H_2O \rightarrow CO_2 + 3H_2 \qquad (10)$$

The hydrogen storage capacity is 18.8 wt.% disregarding the mass of the water. The process is fast at 230-250°C with a suitable catalyst, and the equilibrium is strongly in favour of hydrogen. The enthalpy of reaction at 250°C is +58.7 kJ/mol CH_3OH or +19.6 kJ/mol H_2 or 8.6 % of LHV of methanol. Prior to reforming, methanol and water must be evaporated and this takes another +75.8 kJ/mol H_2 or 11.1 % of LHV of methanol. The total minimum requirement is then 19.7 % of LHV of methanol.
However, the LHV of the produced hydrogen (725.4 kJ/mol $3H_2$) is slightly higher than that of the methanol (685.5 kJ/mol CH_3OH). Taking this upgrading of 39.9 kJ into consideration, the expense for reforming is only 58.7 + 75.8 - 39.9 = 94.6 kJ/mol CH_3OH, which is only 13.8 % of LHV of CH_3OH. Moreover, the energy for evaporation can be taken from the waste heat of a fuel cell provided it is operated above 100°C. In that case, only 2.7 % of LHV is needed for fuel processing. This should be easily obtained from the excess stoichiometry assuming λ =1.2.

9.2 Ammonia (NH3)

Ammonia is sometimes considered as an attractive onboard hydrogen carrier because of its high hydrogen content of 16.6 wt.%, the absence of carbon, and the easy storage. At room temperature, its vapour pressure is less than 10 bar and consequently it can be stored as a liquid at moderate pressure. The major drawbacks are its chemical properties and its stability. It is corrosive and poisonous. As a base, it reacts with acids and it is therefore considered a poison to PEM fuel cells because it reacts with the perfluorosulphonic acid membrane even at levels of 10 ppm (Halseid et al., 2006). Solid oxide fuel cells are able to run on ammonia.

The process of ammonia splitting is endothermic:

$$2NH_3 \rightarrow N_2 + 3H_2 \quad (11)$$
$$+30.6 \text{ kJ/mol } H_2 \text{ or } 12.7 \text{ \% of LHV}$$

and thus high temperatures and low pressure favours hydrogen formation.

If the pressure is set to 1 bar and only a few pct. ammonia are accepted for a subsequent clean-up process, then the temperature must be at least 300-400°C, and the reaction heat must be supplied at that temperature (practically the temperature can not be lower for kinetic reasons either).

Ammonia synthesis is exothermic, and in principle, the heat produced can be utilized. Today however, ammonia is manufactured from natural gas and nitrogen from the air, and the plants are consuming energy. A minimum energy required for synthesis is not estimated. Normally, it is manufactured from natural gas instead of primary hydrogen.

One approach addressing the safety issue is to store ammonia as a complex with a salt, e.g., $MgCl_2$. This idea was recently presented as hydrogen storage in tablet form (Christensen et al., 2005). Dry $MgCl_2$ can reversibly take up 6 molecules of ammonia, and the vapour pressure of ammonia becomes many orders of magnitude smaller. Moreover, in contrast to the hydrides, the complex can be stored in air with only a slow liberation of ammonia. The complex contains 9.1 wt.% hydrogen. Liberation of ammonia is endothermic and the enthalpy for the process is +43 kJ/mol H_2 (Christensen et al., 2005) or 17.8% of LHV. When the enthalpy for ammonia splitting is added, the overall minimum energy is +75 kJ/mol H_2 or 30.5 % of LHV. Liberation of all NH_3 requires a temperature of 350°C, although 2/3 of the ammonia is liberated at 200°C. The system is reversible with respect to ammonia storage, but not with respect to hydrogen storage. Ammonia can be charged onboard, but hydrogen cannot, and ammonia must be synthesized in a plant.

10. Heat available from fuel cells

Fuel cell systems operate with different efficiencies, but in most cases, at least 50 % of the fuel energy is liberated as heat due to different losses, mainly in the fuel cell. This corresponds to 120.9 kJ/mol H_2 (LHV) and is plentiful for any of the storage systems discussed here, even with large transfer losses. The determining factor is the working temperature of the fuel cell (or the exhaust temperature of the combustion engine) because the heat must be delivered at the working temperature of the storage device. Today the PEM fuel cell is almost exclusively considered for vehicles, due to, among other things, the low working temperature of around 80°C. This advantage becomes a disadvantage when heat is needed at higher temperatures, but today there are PEM fuel cells with working temperatures up to 200°C (Li et al., 2009).

11. Conclusion /Comparison

In Table 2, the maximum hydrogen storage efficiencies of the different storage systems are compared. For each system, the onboard hydrogen content is arbitrarily set to 100%.

The column "Energy available onboard" shall be understood as follows: The first percentage listed is the amount available when only heat at 25°C can be used for hydrogen liberation. The value in brackets (when relevant) is the percentage available if a heat source at higher temperature is available. If heat is required, but not available at the working temperature, fuel must be burned to provide it, thus limiting the practical storage capacity.
The column "Energy used for storage process" lists energy consumption for storing the hydrogen (e.g. for compression or charging a hydride). This is in principle an off board energy consumption that does not affect the amount of hydrogen stored. The two sub-columns contain minimum theoretical values and estimated practical values.
The column "Round trip efficiency" gives the overall efficiency for storing and liberating hydrogen.
In all cases energy consumptions due to thermal losses, pumping and the like are omitted.

System	Energy available onboard (% of stored H_2)	Energy used for storage process (% of LHV of stored H_2)		Round trip efficiency (based on the assumptions made)	
	At 25°C	Theoretical minimum	Practical	Theoretical minimum	Practical
Compressed 200 bar	100%	5.4%	10.0%	94.9%	91%
Compressed 800 bar	100%	7.3%	15.5%	93.2%	87%
Liquid hydrogen	~100%	11.8%	40%	89.4%	71%
Interstitial hydride	88% (100% @ ~50°C)	3%	5%	85.4% (97% at 50°C)	84% (95% at 50°C)
MgH_2	69.2% (100% @ 300°C)	0%	5%	69.2% (100% @ 300°C)	66% (95% @ 300°C)
Mg_2NiH_4	73.3% (100% @ 255°C)	0%	5%	73.3% (100% @ 255°C)	70% (95% @ 255°C)
$NaAlH_4$ (3H available)	83.5% (100% @ 150°C)	4.8%	8%	79.7% (95% @ 150°C)	77% (93% @ 150°C)
$NaBH_4$ + water	100%	22%	Very much	82%	Very low
Methanol	81.4% (92.5% @100°C) (100% @ 250°C)	-	-	-	-
Ammonia	87.3% (100% @ 350°C)	-	-	-	-
$Mg(NH_3)_6Cl_2$	69.5% (100% @ 350°C)	-	-	-	-

Table 2. A summary of the efficiencies for storage and release of hydrogen. Note that heat transfer losses and pumping might reduce the practical energy efficiency further.

12. Acknowledgements

The authors wish to thank Nordic Energy Research Programme and Danish Energy Authority for the financial support of the present work.

13. References

Bogdanovic, B.; Brand, R. A.; Marjanovic, A.; Schwickardi, M. & Tolle, J (2000). Metal-doped sodium aluminium hydrides as potential new hydrogen storage materials. *J. Alloys Comp.* 302, 36-58.

Bossel, U.; Eliasson, B. & Taylor, G. (2003). The Future of the Hydrogen Economy: Bright or Bleak? Version of 15. April 2003 updated for distribution at the 2003 Fuel Cell Seminar 3 - 7 November 2003. www.efcf.com/reports.

Christensen, C. H.; Sørensen, R. Z.; Johannessen, T. ; Quaade, U. J.; Honkala, K.; Elmoe, T. D.; Køhler, R. and Jens K. Nørskov (2005). Metal ammine complexes for hydrogen storage. *J. Mater. Chem.* 15, 4106–4108

Halseid, R.; Vie, P. J. S. & Tunold, R. (2006). Effect of ammonia on the performance of polymer electrolyte membrane fuel cells. *J. Power Sources* 154, 343–350.

Jensen, J. O.; Vestbø, A. P.; Li, Q. & Bjerrum, N. J. (2007). The Energy Efficiency of Onboard Hydrogen Storage. *J. Alloys Comp.* 446–447 723–728

Leon, A. (2008). Hydrogen technology. Mobile and portable applications. Springer, ISBN: 978-3-540-79027-3, Berlin, Heidelberg.

Li, Q; Jensen, J. O.; Savinell, R. F. & Bjerrum, N. J. (2009). Acid-doped polybenzimidazole (PBI) membranes for high temperature proton exchange membrane fuel cells. *Prog. Polymer Sci.* 34, 449-477.

Li, Z. P.; Liu, B. H.; Arai, K.; Morigazaki, N. & Suda, S. (2003). Protide compounds in hydrogen storage systems. *J. Alloys Comp.* 356-357 469-474.

Perry, R. H. & Green, D. W. (1984). Perry's Chemical Engineers' Handbook, 6. ed. Mc.Graw-Hill.

Weindorf, W.; Bünger, U. & Schindler, J. (2003). Comments on the paper by Baldur Eliasson and Ulf Bossel "The Future of the Hydrogen Economy: Bright or Bleak?" L-B-Systemtechnik GmbH. On www.hyweb.de 15-07-2003.

Züttel, A.; Borgschulte, A. & Schlapbach, L. (2008). Hydrogen as an energy carrier. Wiley, ISBN: 978-3-527-30817-0, Weinheim.

9

Energy efficiency of Fuel Processor – PEM Fuel Cell systems

Lucia Salemme, Laura Menna and Marino Simeone
University of Naples "Federico II", Department of Chemical Engineering
Italy

1. Introduction

As the world moved into the 21st century, a rapid development in industrial and transportation sectors and improvements in living standards have been observed, leading to a strong growth in the energy demand and in global emissions (Song, 2002). In this context, fuel cell technology has been receiving an increasing attention, thanks to its lower emissions and potentially higher energy efficiency if compared with internal combustion engines. A fuel cell is defined as an electrochemical device in which the chemical energy stored in a fuel is converted directly into electricity. Among all fuel cells, low temperature Proton Exchange Membrane Fuel Cells (PEMFC) are promising devices for decentralized energy production, both in stationary and automotive field, thanks to high compactness, low weight (high power-to-weight ratio), high modularity and efficiency, fast start-up and response to load changes.

The ideal fuel for PEMFC is hydrogen with low carbon monoxide content to avoid poisoning of the fuel cell; in this way, PEMFC can achieve efficiency up to 60%, far higher if compared to 20-35% efficiency of an internal combustion engine.

Hydrogen, though, is not a primary source. It is substantially an energy carrier, that can be stored, transported and used as gaseous fuel, but, it needs to be produced from other fuels.

Today most of the hydrogen produced is obtained by hydrocarbons in large industrial plants through the well-known Steam Reforming and Autothermal Reforming processes.

However, hydrogen distribution from industrial production plants to small-scale users meets some limitations related to difficulties in hydrogen storage and transport. For its chemical and physical properties, indeed, the development of an hydrogen infrastructure seems to be not feasible in short term, while more reasonable seems to be the concept of decentralized hydrogen production; in this way, an hydrogen source, such as methane, is distributed through pipelines to the small-scale plant, placed nearby users, and the *in situ* produced hydrogen is fed directly to the energy production system, avoiding hydrogen storage and transportation. In this sense, a compact fuel processor, capable of generating a hydrogen rich stream from an easily transportable fuel, is a potential root to accelerate PEMFC deployment in the near future.

A typical fuel processor is constituted by a reforming unit coupled with a CO clean-up section, introduced to guarantee hydrogen production with a CO content compatible with

PEMFC specifics. In particular, two different kinds of fuel processor are most frequently described in the scientific literature; a conventional one, in which the reforming unit is followed by two water gas shift (WGS) reactors and a preferential CO oxidation (PROX) reactor (Ersoz et al, 2006), and an innovative one, in which the reforming unit is coupled with highly selective hydrogen membranes to produce pure hydrogen, allowing to operate the PEMFC without a purge stream, generally named as anode off-gas (Lattner et al, 2004).
The global energy efficiency of these systems strictly depends on fuel processor configuration and on operating conditions; therefore, a comprehensive simulative analysis of fuel processors coupled with a PEMFC can contribute to identify the conditions that maximize system performance.
The following paragraphs provide a detailed description of conventional and membrane-based fuel processors. In particular, section 2.1 describes the conventional fuel processors, with details on the reforming technologies and on the typical CO clean-up techniques, while section 2.2 describes innovative fuel processor and membrane technology. Section 2.3 reviews the state of art of the analysis of fuel processor - PEMFC system. Section 3 and 4 report the methodology employed to simulate system performance and the results obtained, respectively. Finally, section 5 draws the main conclusions on the energy efficiency analysis presented.

2. Fuel Processor - PEMFC systems

2.1 Conventional Fuel Processors

Fig. 1 shows the scheme of a conventional fuel processor for hydrogen production from methane, which consists of a desulfurization unit (Des), a syngas production section and a CO clean-up section.

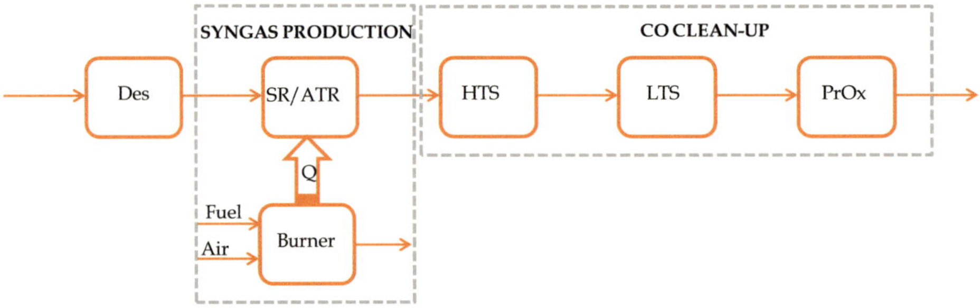

Fig. 1. Conventional Fuel Processor

The desulfurization section is required to lower the sulfur content of the fuel to 0.2 ppm, both for environmental and catalysts restrictions; it generally consists of an hydrodesulphurization reactor, where hydrogen added to the fuel reacts with the sulfur compounds to form H_2S, followed by an adsorption bed to remove H_2S.
The desulfurization process is a quite mature technology and its optimization is essentially related to the catalytic system and it will not be analyzed further. A comprehensive treatment of this unit can be found in Lampert et al, 2004.
The syngas production section is generally used to convert the fuel into syngas, a mixture of H_2 and CO. Two main syngas production technologies are generally employed: Steam

Reforming and Autothermal Reforming. The thermodynamic analysis of reforming processes is widely discussed in the literature (Seo et al, 2002), as well as the optimization of catalyst formulation and operating conditions that maximize process performance (Xu et al, 2006, Semelsberger et al, 2004).
The Steam Reforming process is realized by feeding methane and steam to a catalytic reactor, where the following reactions take place:
1) $CH_4 + H_2O = CO + 3H_2$ $\Delta H^o_R = 49$ Kcal/mol CH_4
2) $CO + H_2O = CO_2 + H_2$ $\Delta H^o_R = -9.8$ Kcal/mol CO
3) $CH_4 = C + 2H_2$ $\Delta H^o_R = 18$ Kcal/ mol CH_4
The operative parameters that influence this process are: pressure (P), temperature (T_{SR}) and steam to methane ratio (H_2O/CH_4) in the feed.
By observing reactions 1, 2 and 3, the reader will be easily convinced that the process occurs with an increment of number of moles; therefore it is favored by low pressures.
The process is globally endothermic and it is favored by high temperatures. The heat required for the reaction is supplied by an external burner fed with additional fuel and air.
Usually, reactor temperature does not exceed 800°C ca. due to catalyst and construction materials constraints.
The value of H_2O/CH_4 employed is usually higher than 2 (stoichiometric value), to reduce coke formation and lower than 4, to limit operative cost and reactor size.
Due to its high selectivity and to the high concentration of hydrogen in the product stream, steam reforming is the most common process to produce hydrogen from hydrocarbons.
However, when looked at from a "decentralized hydrogen production" perspective, it shows some disadvantages essentially because of reduced compactness and slow response to load changes. Both aspects should be attributed to the endothermicity of the reaction and to the high residence times required.
Auto thermal Reforming is obtained by adding air to the inlet SR mixture; in this way, the heat for the endothermic reforming reactions is supplied by the oxidation of part of the methane inside the reactor itself.
The amount of air must be such that the energy generated by the oxidation reactions balances the energy requirement of the reforming reaction, maintaining reactor temperature to typical SR values (600-800°C).
The internal heat generation offers advantages in terms of reactor size and start up times; however, the addition of air to the feed lowers hydrogen concentration in the reformate stream due to the presence of large amounts of nitrogen, fed to the reactor as air.
Either through Steam reforming or Autothermal reforming, the outlet of the reactor has potential of further hydrogen production. Indeed, being reaction 2 exothermic, it is limited by the high temperatures typical of the reforming reactor. For this reason, another reaction step usually follows the main reforming reactor and reduces CO content to less than 10 ppm. This CO clean-up section is constituted by two water gas shift (WGS) reactors and a preferential CO oxidation (PROX) reactor.
The WGS process is a well-known technology, where the following reaction takes place:
4) $CO + H_2O = CO_2 + H_2$ $\Delta H^o_R = -9.8$ Kcal/mol CO
WGS is realized in two stages with inter-cooling; the first stage is generally operated at 350-420°C and is referred to as "high temperature stage" (HTS), whereas the second stage is operated at 200-220°C and is referred to as "low temperature stage" (LTS). The outlet CO concentration from LTS is 0.2 - 0.5% ca. and a further CO conversion stage must be present

before the mixture can be fed to a PEMFC. In conventional fuel processors, CO is reduced to less than 50 ppm in a preferential CO Oxidation (PrOx) stage. The reactor is generally adiabatic and catalyst as well as operating conditions must be carefully chosen, in order to promote CO conversion whilst keeping hydrogen oxidation limited. This CO purification technology is mature and well defined, although it has disadvantage in terms of compactness and catalyst deactivation.
The stream leaving the fuel processor is generally named as reformate and contains the hydrogen produced, as well as CO_2, H_2O, unreacted CH_4 and N_2. This stream is ready to be sent to a PEMFC.

2.2 Innovative Fuel Processors

Innovative Fuel Processors are characterized by the employment of a membrane reactor, in which a high selective hydrogen separation membrane is coupled with a catalytic reactor to produce pure hydrogen.
A typical membrane reactor is constituted by two co-axial tubes, with the internal one being the hydrogen separation membrane; generally, the reaction happens in the annulus and the permeate hydrogen flows in the inner tube.
The stream leaving the reaction is named retentate and the stream permeated through the membrane is named permeate.
Membrane reactor is illustrated in Fig. 2 for the following generic reaction:
$A + B = C + H_2$
The membrane continuously removes the H_2 produced in the reaction zone, thus shifting the chemical equilibrium towards the products; this allows obtaining higher conversions of reactants to hydrogen with respect to a conventional reactor, working in the same operating conditions.
A typical membrane used to separate hydrogen from a gas mixture is a Palladium or a Palladium alloy membrane (Shu et al., 1991); this kind of membrane is able to separate hydrogen with selectivity close to 100%. Hydrogen permeation through Palladium membranes happens according to a solution/diffusion mechanism and the hydrogen flux through the membrane, J_{H2} is described by the following law:

$$J_{H2} = \frac{\mu_{H2} A}{\delta}\left(\sqrt{P_{H2,R}} - \sqrt{P_{H2,P}}\right) \qquad (1)$$

where μ_{H2} is the permeability coefficient [mol/(m^2 s Pa$^{0.5}$)], A is the membrane surface area [m^2], δ is the membrane thickness [m] and $P_{H2,R}$ and $P_{H2,P}$ are hydrogen partial pressures [kPa] on the retentate side and on the permeate side of the membrane, respectively. Eq. 1 is known as Sievert's law and it is valid if the bulk phase diffusion of atomic hydrogen is the rate limiting step in the hydrogen permeation process.
To increase the separation driving force, usually the retentate is kept at higher pressure than the permeate. In common applications, permeate pressure is atmospheric and retentate pressure is in the range 10-15 atm (compatibly with mechanical constraints).
A possible way to further increase the separation driving force is to reduce hydrogen partial pressure in the permeate ($P_{H2,P}$) by diluting the permeate stream with sweep gas (usually superheated steam).

Sievert's law shows that an increase of the hydrogen flux is achieved with reducing membrane thickness. Palladium membranes should not be far thinner than 80-100 μm due to mechanical stability of the layer and to the presence of defects and pinholes that reduce hydrogen selectivity. To overcome this problem, current technologies foresee a thin layer (20-50 μm) of Pd deposited on a porous ceramic or metal substrate.
Another important issue of Pd membranes (pure or supported) is thermal resistance. Temperature should not be less than 200°C, to prevent hydrogen embrittlement and not higher than 600°C ca. to prevent material damage.

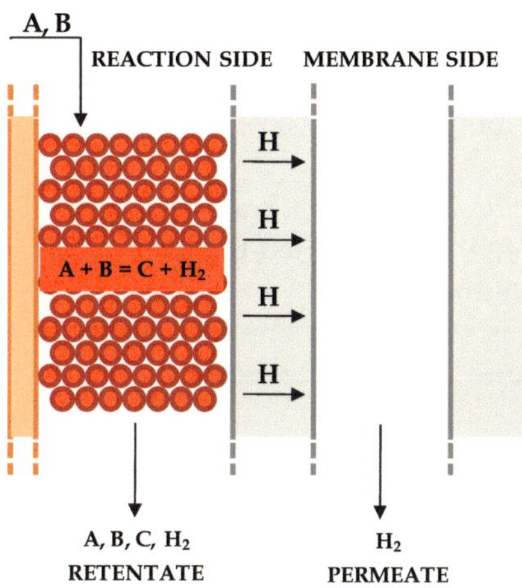

Fig. 2. Membrane Reactor

Innovative fuel processors can be realized by combining the membrane either with the reforming unit, generating the fuel processor reported in Fig. 3 (FP.1), or with a water gas shift unit, generating the fuel processor reported in Fig. 3 (FP.2).

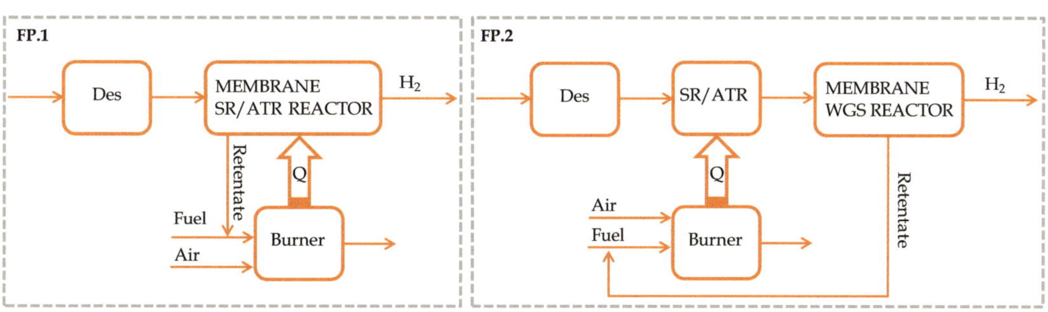

Fig. 3. Innovative Fuel Processors

FP.1 consists of a desulfurization unit followed by a membrane reforming reactor, with a burner. This solution guarantees the highest compactness in terms of number of units, since it allows to totally suppress the CO clean-up section; indeed, when the membrane is integrated in the reforming reactor, the permeate stream is pure hydrogen, that can be directly fed to a PEMFC.

However, this solution limits the choice of the operating temperature of the process that must be compatible with the constraints imposed by the presence of a membrane.
FP.2 consists of a desulfurization unit followed by a reforming reactor and a membrane water gas shift reactor. In this case, the membrane is placed in the low temperature zone of the fuel processor, operating at thermal levels compatible with its stability. This solution, although less compact than the previous one, allows to operate the syngas production section at higher temperature.

2.3 PEMFC

A fuel cell is an electrochemical device that converts the chemical energy of a fuel directly into electrical energy. Intermediate conversions of the fuel to thermal and mechanical energy are not required. All fuel cells consist of two electrodes (anode and cathode) and an electrolyte.
Proton exchange membrane fuel cells, also known as polymer electrolyte membrane fuel cells (PEMFC), are a type of fuel cell in which the electrolyte is a polymeric membrane and the electrodes are made of platinum.
In a PEMFC unit, hydrogen is supplied at one side of the membrane where it is split into hydrogen protons and electrons, at anode electrode:

$H_2 \rightarrow 2H^+ + 2e^-$

The protons permeate through the polymeric membrane to the reach the cathode electrode, where oxygen is supplied and the following reactions takes place.

$O_2 + 4H^+ + 4e^- \rightarrow 2H_2O$

Electrons circulate in an external electric circuit under a potential difference.
The electric potential generated in a single unit is about 0.9V. To achieve a higher voltage, several membrane units need to be connected in series, forming a fuel cell stack. The electrical power output of the fuel cell is about 60% of its energy generation, the remaining energy is released as heat.
Generally, oxygen is fed to the cathode as an air stream; in practical systems, an excess of oxygen is fed to the cathode to avoid extremely low concentration at the exit. Frequently, a 50% or higher excess with respect to the stoichiometric oxygen is fed to the cathode.
For the anode, instead, it is not typically the stoichiometric ratio, but rather the amount of hydrogen converted to the fuel cell as a percentage of the feed that is specified. This amount is named as the hydrogen utilization factor U_f; when pure hydrogen is fed to the PEMFC, this factor can be assumed equal to unity.
For PEMFC systems running on reformate produced in a conventional fuel processor, this factor can be assumed equal to 0.8. This implies that not all gas fed to the anode is converted and unconverted hydrogen and the rest of the reformate is purged off as a stream named as Anode Off-Gas (AOG). This stream presents a heating value due to the presence of hydrogen and methane; therefore, it can be used in the burner of the conventional fuel processor to eventually supply heat to the process.

2.4 System Analysis of Fuel Processor - PEMFC systems

Optimization of energy efficiency of a fuel processor PEMFC system is a central issue in actual research studies. Since the efficiency of the PEMFC can be assumed as a constant equal to 60%, the efficiency of the entire system depends on fuel processor efficiency and on the integration between the fuel processor and the PEMFC.

The optimization of system efficiency is achieved by exploring the effect of the operating parameters considering, at the same time, the heat recovery between the various streams and units present in the system and the necessary driving force for heat exchange.

The optimization of conventional hydrocarbon-based fuel processors has been tackled by several authors who have identified the most favorable operating conditions to maximize the reforming efficiency. As a general outcome, SR-based fuel processors provide the highest hydrogen concentration in the product stream, whereas the highest reforming efficiency is reached with ATR-based fuel processors, due to the energy loss represented by the latent heat of vaporization of the water that escapes with the combustion products in the SR system (Ahmed et al, 2001).

However, as the system grows in complexity, due to the presence of the fuel cell, optimization of the global energy efficiency must also take into account the recovery of the energy contained in the spent gas released at the cell anode (anode off-gas). Ersoz et al. (2006) performed an analysis of global energy efficiency on a fuel processor - PEMFC system, considering methane as the fuel and steam reforming, partial oxidation and auto thermal reforming as alternative processes to produce hydrogen. Their main conclusion is that the highest global energy efficiency is reached when SR is used, essentially due to the higher recovery of anode off-gas heating value.

As far as membrane-based fuel processor is concerned, only few contributions which address the behavior of the entire system are available, that include not only the membrane-based fuel processor, but also the fuel cell, the auxiliary power units and the heat exchangers (Pearlman et al, Lattner et al, Manzolini et al, Campanari et al, Lyubovsky et al). Most of these studies refer to liquid fuels and only few contributions are available when methane is employed.

In particular, Campanari et al. (2008) analyzed an integrated membrane SR reactor coupled with a PEMFC, showing that a higher global energy efficiency can be achieved, with respect to conventional fuel processors, if a membrane reactor is employed.

Lyubovsky et al. (2006) analyzed a methane ATR-based fuel processor - PEMFC system, with a membrane unit placed downstream the WGS unit and operating at high pressure, concluding that high global energy efficiency can be obtained if a turbine is introduced in the system to generate additional power from the expansion of the hot gases produced by the combustion of the membrane retentate stream.

In order to have a complete vision of the effect of system configuration and of operating parameters on the efficiency of fuel processor - PEMFC systems, a comprehensive analysis of different configurations will be presented and compared in terms of energy efficiency; in particular, methane will be considered as fuel and SR and ATR as reforming processes; the focus of the discussion will be about the following fuel processor (FP) configurations, each coupled with a PEMFC:

FP.A) SR reactor, followed by two WGS reactors and a PROX reactor.
FP.B) ATR reactor, followed by two WGS reactors and a PROX reactor.
FP.C) Integrated membrane-SR reactor.
FP.D) Integrated membrane-ATR reactor.
FP.E) SR reactor followed by a membrane WGS reactor.
FP.F) ATR reactor followed by a membrane WGS reactor.

Each system configuration is investigated by varying operating parameters, such as steam to methane and oxygen to methane inlet ratios, reforming temperature, as well as pressure; the effect of the addition of steam as sweep gas on the permeate side of the membrane reactors will be also presented and discussed.

3. Methodology

The simulations were performed in stationary conditions, by using the commercial package Aspen Plus®. The selected property method was Peng-Robinson and the component list was restricted to CH_4, O_2, N_2, H_2O, CO, H_2 and CO_2.
Methane was considered as fuel, fed at 25°C and 1 atm, with a constant flow rate of 1 kmol/h. Feed to the system was completed with a liquid water stream (25°C and 1 atm) both in SR and ATR-based FPs; an air stream (25°C and 1 atm) is also present in the ATR-based FPs.
The configurations simulated (flow sheets) are presented in the following sections, where the assumptions and the model libraries used to simulate the process are presented. Section 3.1 is dedicated to conventional fuel processors, whereas membrane-based fuel processors are described in section 3.2. The quantities employed to calculate energy efficiency are defined in section 3.4.

3.1 Conventional fuel processor – PEMFC systems

Fig. 4 reports the flow sheet of a conventional SR-based fuel processor coupled with a PEMFC (FP.A). The fuel processor consists of a reforming and a CO clean-up section.

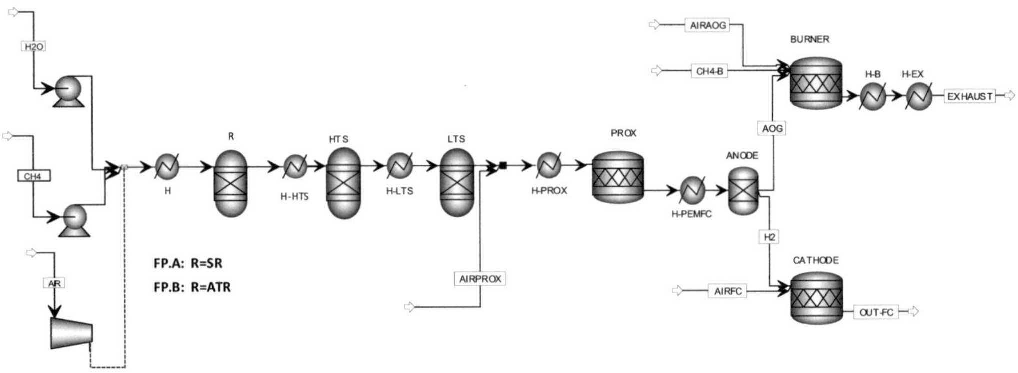

Fig. 4. Flowsheet of fuel processor FP.A and FP.B coupled with a PEMFC

The reforming section is an isothermal reactor (SR), modeled by using the model library ***RGIBBS.***
The CO clean-up section consists of a high (HTS) and low (LTS) temperature water gas shift reactor followed by a PROX reactor. HTS and LTS were modeled by using model library ***RGIBBS***; the reactors were considered as adiabatic and methane was considered as an inert in order to eliminate the undesired methanation reaction, kinetically suppressed on a real catalytic system.

The inlet temperature to the HTS reactor was fixed at 350°C, while the inlet temperature to the LTS one at 200°C. The PROX reactor was modeled as an adiabatic stoichiometric reactor, ***RSTOIC***; this kind of reactor models a stoichiometric reactor with specified reaction extent or conversion; in the case of PROX, two reactions were considered: oxidation of CO to CO_2 with complete conversion of CO and oxidation of H_2 to H_2O; the air fed to the PROX reactor (AIR-PROX) was calculated in order to achieve a 50% oxygen excess with respect to the stoichiometric amount required to convert all the CO to CO_2. The ***RSTOIC*** specifics were completed with the assignment of total conversion of CO and O_2. The inlet temperature to the PROX reactor was fixed at 90°C.
The PEM fuel cell section is simulated as the sequence of the anode, modeled as an ideal separator, ***SEP***, and the cathode, modeled as an isothermal stoichiometric reactor, ***RSTOIC.*** The presence of the ***SEP*** unit allows to model a purge gas (anode off-gas, AOG) required for mass balance reasons, whenever the hydrogen stream sent to the PEM fuel cell is not 100% pure. In agreement with the literature, the hydrogen split fraction in the stream H2 at the outlet of the ***SEP*** was fixed at 0.75 (Francesconi et al, 2007), whereas the split fractions of all the other components were taken as 0. The ***RSTOIC*** unit models the hydrogen oxidation reaction occurring in the fuel cell. The reactor specifics were completed by considering an operating temperature of 80°C and pressure of 1 atm; the inlet air at the cathode (AIR-FC), fed at 25°C and 1 atm, guarantees a 50% excess of oxygen in the ***RSTOIC*** reactor. In agreement with Ratnamala et al (2005), these conditions were considered as sufficient to assign total hydrogen conversion. The anode off-gas is sent to a burner, modeled as an adiabatic ***RSTOIC***, working at atmospheric pressure with 50% excess air (AIR-B); the complete combustion of all fuels contained (i.e. hydrogen, methane, carbon monoxide) was always imposed.
The heat required by the SR reactor working at temperature T_{SR} is supplied by the heat exchanger H-B, where the stream coming from the burner is cooled to T_{SR} +10°C. Model library ***HEATER*** was used for this purpose. An additional stream of fuel (CH4-B) is sent to the burner to satisfy the global heat demand of the system, when needed. The heat removed for cooling the stream at the outlet of heat exchanger H-B, at the inlet of HTS, LTS and PROX reactors and PEM fuel cell, as well as the heat for keeping the PROX at constant temperature, is employed to preheat the SR inlet stream. On the other hand, the heat removed for cooling the PEM fuel cell, is not recovered, since most of the times a simple air fan is used to cool the stack.
As concerns the flow sheet of a conventional ATR-based fuel processor (FP.B) coupled with a PEM fuel cell, for the sake of simplicity, the description of the flow sheet will be carried out by indicating the differences with respect to the flow sheet of Fig. 4, which are concentrated only in the reforming section. Indeed, in FP.B the reforming section is constituted by an adiabatic reactor (ATR), modeled by using model library ***RGIBBS***. The heat exchanger H-B can be suppressed in this configuration, since the ATR reactor has no heat requirement. The inlet temperature to the ATR reactor is fixed at 350°C, and is regulated by means of the heat exchanger H-ATR.

3.2 Innovative fuel processor – PEMFC systems

The integrated membrane-reactors were simulated by discretizing the membrane reactor with a series of N reactor-separator units. With this approximation, reactors are assumed to reach equilibrium and the separators are modeled as ideal separators, ***SEP***, whose output is given by a stream of pure hydrogen (permeate) and a stream containing the unseparated hydrogen and all the balance (retentate). The amount of hydrogen separated ($n^{i}_{H2,P}$) is calculated assuming equilibrium between the partial pressure in the retentate and permeate side, according to Eq.2:

$$P_R \cdot \frac{n^{i}_{H2,R} - n^{i}_{H2,P}}{n_{R,i} + n^{i}_{H2,P}} = P^{i}_{H2,P} \tag{2}$$

where P_R is the pressure in the retentate side of the membrane, equals to reactor pressure; $n^{i}_{H2,R}$ is the mole flow of hydrogen in the retentate stream; n^{i}_{R} is the total mole flow of the retentate stream; $P^{i}_{H2,P}$ is hydrogen partial pressure in the permeate side of the membrane, calculated as:

$$P^{i}_{H2,P} = \frac{n^{i}_{H2,P}}{n^{i}_{H2,P} + n_{SG}} \cdot P_P \tag{3}$$

where P_P is the pressure in the permeate side of the membrane, taken as 1 atm in all the simulations, and n_{SG} represents the molar flow rate of steam sweep gas (SG), which can be introduced to increase the separation driving force in the membrane. When present, the sweep gas is produced by liquid water, fed at 25°C and 1 atm to a heat exchanger and sent to the membrane reactor in countercurrent flow mode.

The high hydrogen purity of the stream sent to the PEMFC allows taking as zero the anode off-gas, simplifying the model of the PEMFC to the cathode side (***RSTOIC***) only.

Fig. 5 reports the flow sheet used to simulate a membrane-SR reactor (FP.C) and a membrane-ATR reactor (FP.D) coupled with a PEMFC. The membrane-SR reactor was discretized with 30 units, whereas the membrane-ATR reactor was discretized with 20 units; the number of units required to model each membrane-reactor was assessed by repeating the simulations with an increasing number of reactor-separator units and was chosen as the minimum value above which global efficiency remained constant within ± 0.1%.

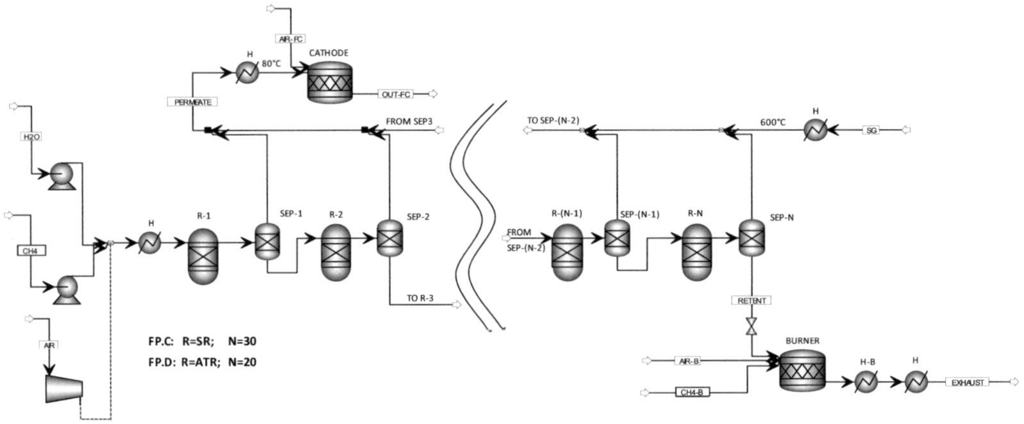

Fig. 5. Flowsheet of fuel processors FP.C and FP.D coupled with a PEMFC

As for the case of the conventional system, the heat eventually required by the reforming reactor is supplied by the heat exchanger H-B and an additional stream of fuel (CH4-B) is sent to the burner to satisfy the global heat demand of the system, when needed. The heat removed for cooling the streams at the outlet of heat exchanger H-B and at the inlet of the PEMFC is recovered to preheat SR inlet stream and eventually to produce sweep gas.
Fig. 6 reports the flow sheet used to simulate a SR-based FP coupled with a PEMFC, where the SR reactor is followed by a membrane WGS reactor (FP.E) and a ATR-based FP coupled with a PEMFC, where the ATR reactor is followed by a membrane WGS reactor (FP.F). With respect to FP.A and FP.B, in this case only one Water Gas Shift reactor is present, with an inlet temperature of 300°C; the membrane WGS reactor was discredited into four units.
As for the case of described above, the heat eventually required by the reforming reactor is supplied by H-B and an additional stream of fuel (CH4-B) is sent to the burner to satisfy the global heat demand of the system, when needed. The heat removed for cooling the streams at the outlet of heat exchanger H-B and at the inlet of the PEM fuel cell is recovered to preheat SR inlet stream and eventually to produce sweep gas.

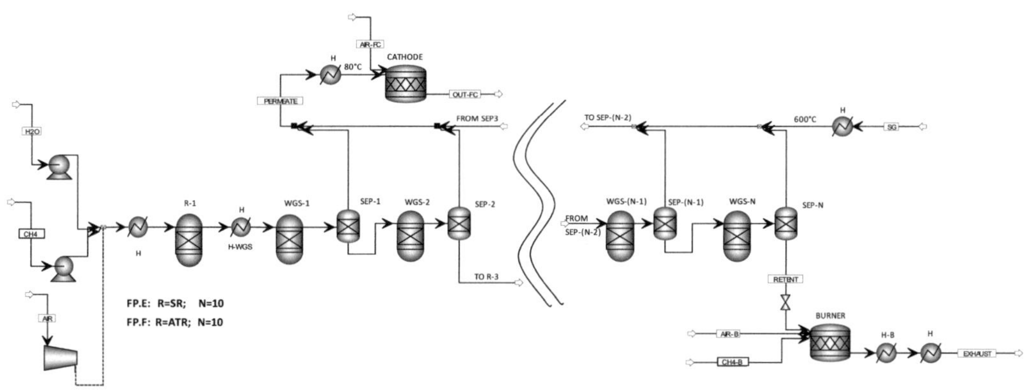

Fig. 6. Flow sheet of fuel processor FP.E and FP.F coupled with a PEMFC

All the reactors were considered as operating at the same pressure.
Auxiliary power units for compression of the reactants were considered in the configurations where pressure was explored as an operation variable, i.e. FP.C and FP.D.
100°C was chosen as the minimum exhaust gas temperature (T_{ex}), when compatible with the constraint of a positive driving force in the heat exchangers present in the plant.
Finally, it is worth mentioning that the assumptions made to model the system are the same for all the configurations investigated and do not affect the conclusions drawn in this comparative analysis.

3.3 System Efficiency

Energy efficiency, η, was defined according to the following Eq.4:

$$\eta = \frac{P_e - P_a}{(n_{CH4,F} + n_{CH4,B}) \cdot LHV_{CH4}} \tag{4}$$

where P_a is the electric power required by the auxiliary units for compression of methane, air and water, $n_{CH4,F}$ is the inlet molar flow rate of methane to the reactor, $n_{CH4,B}$ is the molar flow rate of methane fed to the burner, LHV_{CH4} is the lower heating value of methane and P_e is the electric power generated by the fuel cell, calculated as:

$$P_e = n_{H2} \cdot LHV_{H2} \cdot \eta_{FC} \tag{5}$$

where n_{H2} is the molar flow rate of hydrogen that reacts in the fuel cell, LHV_{H2} is the lower heating value of hydrogen, η_{FC} is the electrochemical efficiency of the cell, taken as 0.6 (Hou et al, 2007). In the membrane-based fuel cell systems, an important parameter is the global hydrogen recovery (HR), defined as:

$$HR = \frac{\sum_{i=1}^{N} n^{i}_{H2,P}}{n_{H2,R} + \sum_{i=1}^{N} n^{i}_{H2,P}} \tag{6}$$

where $n^{i}_{H2,P}$ is the molar flow rate of hydrogen separated by the i-th membrane unit, $n_{H2,R}$ is the molar flow rate of hydrogen in the RETENT stream at the exit of the last separator and N is the number of separators.

According to the definitions given above, η can be expressed as it follows:

$$\eta = (HR \cdot \eta_R \cdot \eta_{FC} - f_a) \cdot (1 - \alpha) \tag{7}$$

where f_a is the fraction of inlet methane required to run the auxiliary units, defined by Eq. 8:

$$f_a = \frac{P_a}{n_{CH4,F} \cdot LHV_{CH4}} \tag{8}$$

α is the ratio between methane flow rate fed to the burner and total methane flow rate fed to the system, defined by Eq. 9:

$$\alpha = \frac{n_{CH4,B}}{n_{CH4,F} + n_{CH4,B}} \tag{9}$$

f_R is the reforming factor, defined by Eq. 10:

$$f_R = \frac{\left(n_{H2,R} + \sum_{i=1}^{N} n^{i}_{H2,P} \right) \cdot LHV_{H2}}{n_{CH4,F} \cdot LHV_{CH4}} \tag{10}$$

This factor is related to the global amount of hydrogen produced in the fuel processor per moles of methane fed to the reforming reactor; therefore it does not depend on the heat requirement of the system.

4. Results

Simulation where performed by varying the main operating parameters for each system. The parameters investigated and the ranges explored are reported in Table 1. For conventional systems (FP.A and FP.B) pressure was fixed at 1 atm since reforming processes are inhibited by pressure increase, whereas the WGS and PROX processes are independent of pressure. The operating ranges of H_2O/CH_4 and T_{SR} for the system with membrane SR reactor (FP.C) are chosen in order to guarantee thermal stability of the membrane and to avoid coke formation. The pressure range investigated for the innovative systems was chosen in order to guarantee the mechanical resistance of the membrane. The operating ranges of H_2O/CH_4 and of O_2/CH_4 for the ATR systems are chosen in order to avoid coke formation and to guarantee the autothermicity of the process (Seo et al, 2002).

	Case	H_2O/CH_4	O_2/CH_4	T_{SR} [°C]	SG/CH_4	P [atm]
SR	FP.A	2.0 – 6.0	-	600 - 800	-	1
	FP.C	2.5 – 6.0	-	500 - 600	0 – 3.0	3 - 15
	FP.E	2.0 – 6.0	-	600 - 800	0 – 3.0	3 - 15
ATR	FP.B	1.2 – 4.0	0.3 – 1.0	-	-	1
	FP.D	1.2 – 4.0	0.3 – 1.0	-	0 – 3.0	3 - 15
	FP.F	1.2 – 4.0	0.3 – 1.0	-	0 – 3.0	3 - 15

Table 1. Range of operating parameters investigated

4.1 Conventional Fuel Processors

Fig. 7 shows the trend of energy efficiency η, methane conversion x_{CH4}, reforming factor f_R and the fraction of total inlet methane that is sent to the burner α as a function of H_2O/CH_4, parametric in the steam reforming reactor temperature.

For all the temperatures investigated, an increase of water content in the feed has a positive effect on methane conversion x_{CH4} and on the reforming factor f_R. This well note trend is due to the fact that water is a reactant of reforming reactions.

For each temperature and until a certain value of H_2O/CH_4, the value of α is equal to zero. For higher H_2O/CH_4, the increase of this ratio leads to an increase of α; indeed, the increase of H_2O/CH_4 causes an increase of the heat required to sustain the reforming process, moreover the improvement of reforming reactor performance with H_2O/CH_4 causes a reduction of the heating value of the AOG stream, thus an increase of the quantity of methane that needs to be sent to the burner for sustaining the endothermicity of the process.

As described in the System efficiency Section, the energy efficiency is a combination of f_R and of α; indeed, η shows a non monotone trend as a function of H_2O/CH_4 because, although an increase of water content causes a continuous increase of reforming reactor performance, the amount of methane sent to the burner also increases with H_2O/CH_4.

For all the H_2O/CH_4 investigated, the increase of reforming reactor temperature (T_{SR}) causes an increase of x_{CH4}, f_R and α. Energy efficiency η shows a different trend on the basis of the weight of these factors: for low H_2O/CH_4, η shows a continuous increase with T_{SR} in the range investigated, whereas, for high H_2O/CH_4, η shows a non monotone trend with T_{SR}.

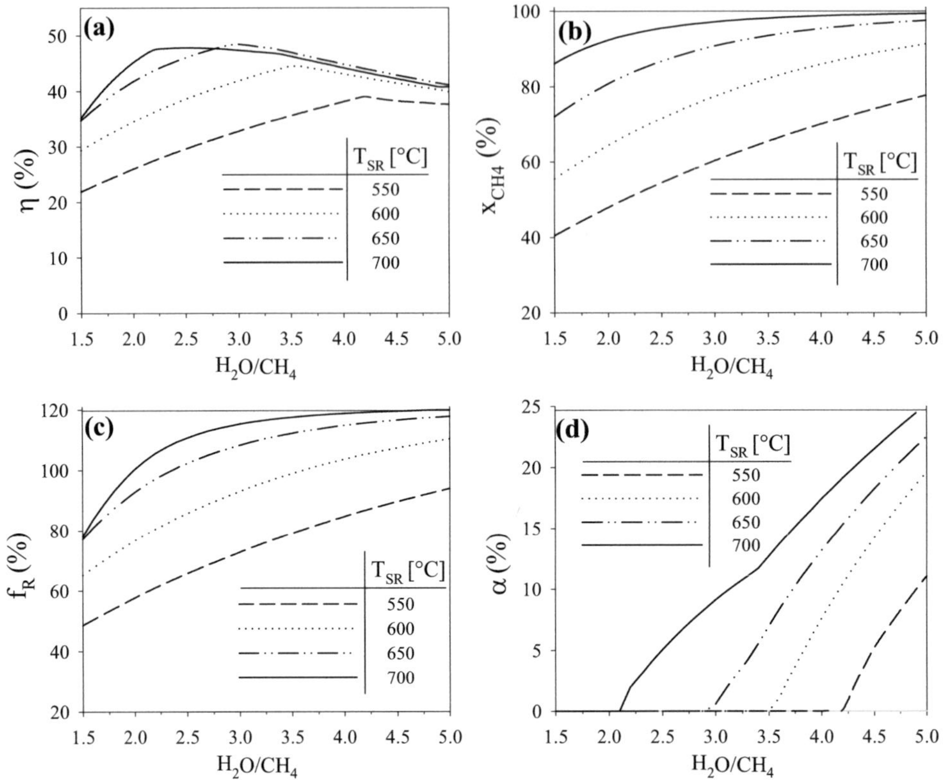

Fig. 7. η (a), x_{CH4} (b), f_R (c) and α (d) as a function of H_2O/CH_4 parametric in T_{SR}

Fig. 8 shows the trend of energy efficiency η, methane conversion x_{CH4}, reforming factor f_R for conventional ATR-based fuel processor - PEMFC systems (systems with FP.B), as a function of O_2/CH_4 parametric in H_2O/CH_4.

Methane conversion shows a monotone increase as a function of O_2/CH_4. The effect of water addition on methane conversion is positive in case x_{CH4} is far lower than unity, whereas this effect can be considered as negligible when the conversion approaches to unity. Reforming factor shows a non monotone trend as a function of O_2/CH_4; indeed, for low O_2/CH_4 values the process cannot reach the temperature values that favor the reforming reactions, whereas for high O_2/CH_4 values, although the reforming temperature results to be strongly increased, the hydrogen and methane oxidation reactions are favorite, with subsequent reduction of the amount of hydrogen produced and, thus, of the f_R.

The addition of water leads to an increase of f_R, being water a reactant of the reforming reactions; this increase becomes negligible for H_2O/CH_4 values higher than 2.

For all the O_2/CH_4 and H_2O/CH_4 values investigated, α remains equal to zero, therefore, the trend of energy efficiency results to be the same of the reforming factor; moreover, there is a waste of heat from the system, related to the autothermic nature of the process, which hinders the possibility of recovering the energy content of the AOG.

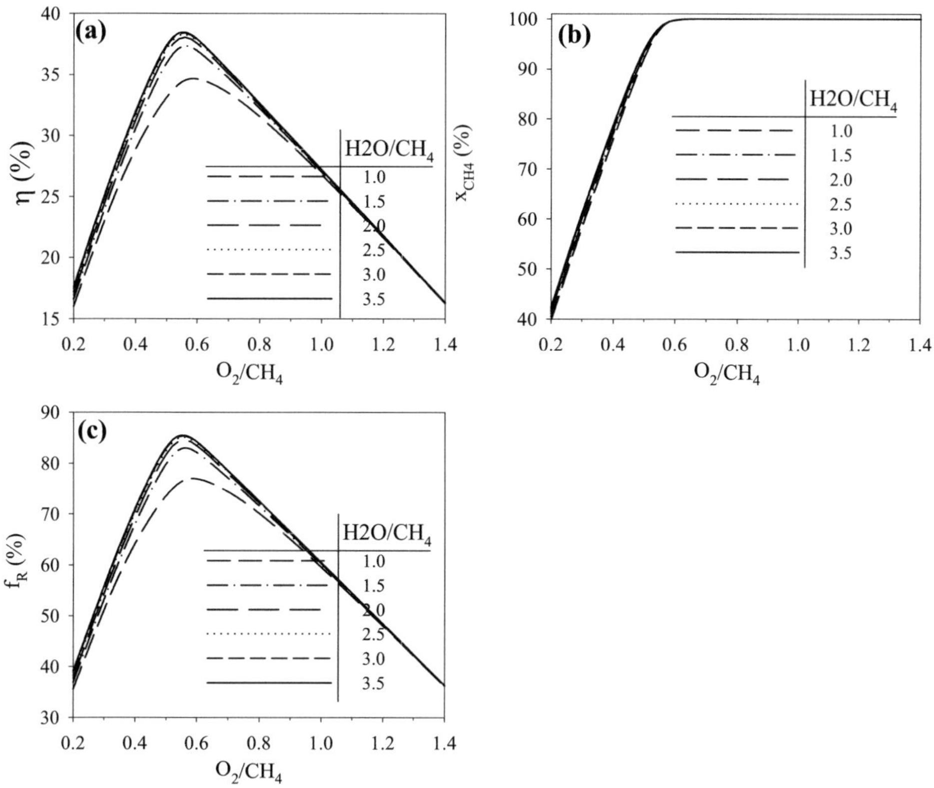

Fig. 8. η (a), x_{CH4} (b) and f_R (c) as a function of O_2/CH_4 parametric in H_2O/CH_4.

Table 2 reports the simulation results and the value of the operative parameters given as simulation input that maximize the energy efficiency η, for FP.A and for FP.B, respectively. FP.A shows the highest global efficiency (48.0%) at T_{SR}=670°C and H_2O/CH_4=2.5. It should be noticed that, in the optimal conditions, methane conversion (x_{CH4}) is lower than unity; however, the non converted methane is not energetically wasted, since it contributes to the energy content of the AOG, used to sustain the endothermicity of the SR reactor. In this conditions, no addition of methane to the burner is needed (α=0). According to the flow sheet of FP.A, the minimum exhaust gas temperature achievable is 226°C. Further heat recovery is hindered by temperature cross-over in the heat exchangers.

Simulation results				
	x_{CH4}	α	η	T_{EX} (°C)
FP.A (SR)	91.0	0.0	48.0	226
FP.B (ATR)	98.8	0.0	38.5	444
Simulation Input				
	P (atm)	**H_2O/CH_4**	**O_2/CH_4**	**T_{SR} (°C)**
FP.A (SR)	1	2.5	-	670
FP.B (ATR)	1	4.0	0.56	-

Table 2. Conventional Fuel Processor – PEMFC systems

FP.B shows the highest global efficiency (38.5%) at O_2/CH_4=0.56 and H_2O/CH_4=4.0; the value of η is significantly lower than what achieved with FP.A, mainly due to the autothermal nature of the ATR process, that limits the possibility to recover the energy content of the AOG. This reflects into a higher exhaust gas temperature in FP.B (444°C) than in FP.A (226°C).

4.2 Innovative Fuel Processors

Fuel Processors based on membrane reforming reactor

Fig. 9 reports the energy efficiency of system with FP.C as a function of pressure.
Energy efficiency rapidly increases with pressure in the range 3-5 atm, where no methane addition to the burner is required to sustain the endothermic steam reforming reaction.

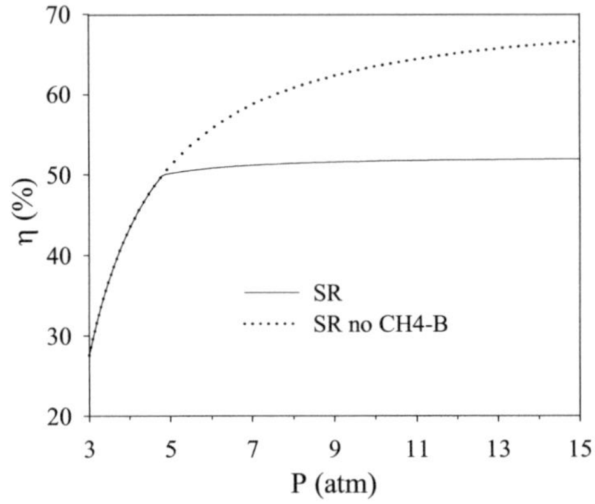

Fig. 9. η as a function of pressure for system with FP.C. Operating conditions: T_{SR}=600°C, H_2O/CH_4=2.5, SG/CH_4=0

As pressure increases above 5 atm ca, η continues to grows with pressure, but at a lower rate, because methane addition to the burner becomes necessary. The dotted line, superimposed to Fig. 9 as an aid to this discussion, represents the value of η that would be calculated if the methane sent to the burner was not factored in the computation.
The trend of η vs P is the combined effect of hydrogen recovery (HR), reforming factor (f_R), the power of the auxiliary units (related to f_a), whose values are reported in Table 3 together with the value of methane conversion (x_{CH4}) and fraction of methane sent to the burner (α).

P (atm)	x_{CH4}	α	T_{EX} (°C)	HR	fa	f_R	η
3	70.6	0	803.8	58	0.5	80.4	27.5
5	86.3	1.8	100	85.9	0.7	100.5	50.2
7	91.8	12.8	100	91.9	0.9	108.0	51.2
9	94.5	17.2	100	94.4	1.1	111.8	51.5
12	96.6	20.4	100	96.2	1.3	114.9	51.8
15	97.6	22	100	97.1	1.4	116.7	51.9

Table 3. System with FP.C. Operating conditions: T_{SR}=600°C, H_2O/CH_4=2.5, SG/CH_4=0

In particular, HR increases with pressure due to the increase of hydrogen separation driving force through the membrane; f_R increases with pressure because it is positively influenced by the trend of HR with pressure, due to the positive effect on reaction equilibrium of increasing hydrogen separation. f_a increases with pressure, due to increasing compression ratios. To complete the picture, it should be kept in mind that the heating value of the retentate decreases with pressure, as a consequence of higher x_{CH4} and HR. This, in turn, influences the quantity of methane sent to the burner to sustain the endothermic steam reforming reaction.
In the low pressure range, the positive effect of HR and f_R on energy efficiency overrules the negative effect of f_a and α. The plateau value reached at higher pressure indicates that the drawback of f_a and α compensates the positive effect of HR and f_R.
Fig. 10 reports the effects of SG/CH_4 on system efficiency of FP.C parametric in pressure, at a fixed outlet exhaust gases temperature of 100°C. Simulation details for P = 10 atm are reported in Table 4.

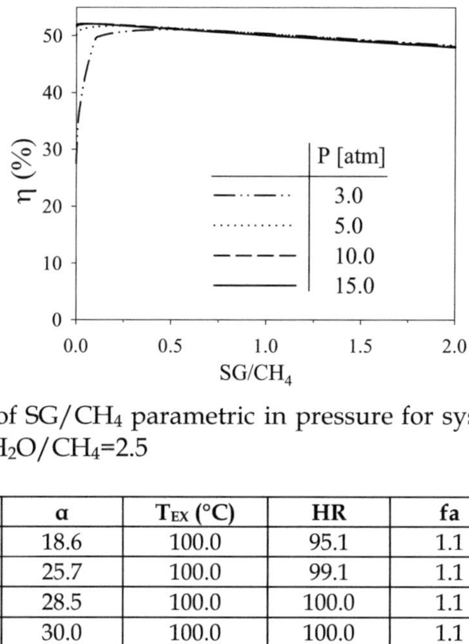

Fig. 10. η as a function of SG/CH_4 parametric in pressure for system with FP.C. Operating conditions: T_{SR}=600°C, H_2O/CH_4=2.5

SG/CH_4	x_{CH4}	α	T_{EX} (°C)	HR	fa	f_R	η
0.0	95.4	18.6	100.0	95.1	1.1	113.1	51.6
0.1	99.8	25.7	100.0	99.1	1.1	119.8	52.1
0.5	100.0	28.5	100.0	100.0	1.1	121.4	51.3
1.0	100.0	30.0	100.0	100.0	1.1	121.3	50.2
1.5	100.0	31.5	100.0	100.0	1.2	121.3	49.1
2.0	100.0	33.0	100.0	100.0	1.2	121.5	48.1

Table 4. System with FP.C. Operating conditions: T_{SR}=600°C, H_2O/CH_4=2.5, P=10 atm

It is possible to observe that η shows a maximum as a function of SG/CH_4 ratio, which shifts leftwards and upwards as pressure increases. For each pressure value investigated, hydrogen recovery is enhanced by the presence of the sweep gas, as a consequence of reduced hydrogen partial pressure in the permeate side; this leads also to an increase of f_R thanks to the positive effect of hydrogen removal on reactions equilibrium.
However, the production of sweep gas is always coupled with addition of methane to the burner, with an increment of α that can overrule the increment of HR and f_R. For this reason,

being η combination of f_R, HR and α, after an initial small increment, it decreases with addition of sweep gas.
The effect of pressure depends on the SG/CH_4 value. For low SG/CH_4, an increase of pressure causes an increase of η, whereas a decreasing trend of the η with pressure is observed at high SG/CH_4. This is due to the fact that the increment of pressure increases both HR and f_a; when SG/CH_4 is high, HR becomes close to 100% already at low pressure values, therefore an increase of pressure only causes an increase of f_a, lowering η.
Table 5 report the detail of the simulation results and value of the operating parameters given as simulation input that maximize the energy efficiency η, for FP.C.
The best way to operate a membrane SR system is to increase the pressure without addition of sweep gas.
It is possible to observe that the energy efficiency of a SR-based system is increased if a membrane reactor is used (FP.C), in place of a conventional reactor. This is due to the possibility to recover a higher amount of heat in FP.C than in FP.A. Indeed the heat exchanger network needed in FP.A has to satisfy the temperature requirements of the Shift and PROX reactors resulting in a higher exhaust gas temperature (226°C), while in FP.C the heat exchanger network allows to cool the exhaust gas to 100°C (as chosen in the methodology), without any temperature cross over.

Simulation results						
	f_R	α	HR	f_a	η	T_{EX} (°C)
FP.C (SR)	120.0	25.6	99.2	1.3	52.2	100.0
Simulation Input						
		P (atm)	H_2O/CH_4	T_{SR} (°C)	SG/CH_4	
FP.C (SR)		15	2.5	600	0.1	

Table 5. Membrane SR – PEMFC system based on membrane reforming reactor

Fig. 11 reports energy efficiency of system with FP.D as a function of pressure. As for the case of FP.C, η shows a continuous increase with pressure, but the values are significantly lower, due to limited recovery of the energy contained in the retentate stream and to the negative contribution of the compressor (see T_{ex} and f_a in Table 6).

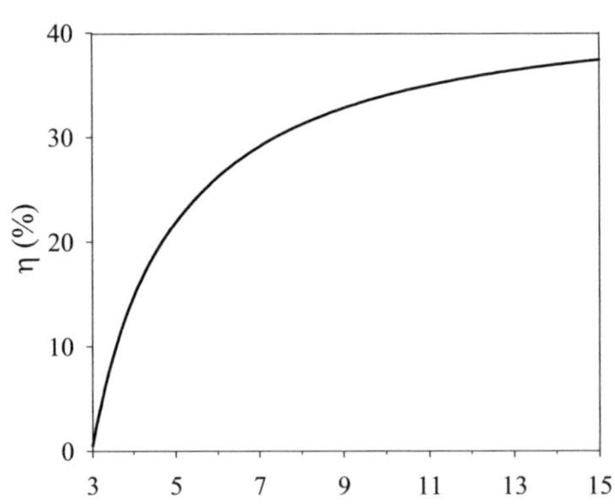

Fig. 11. η as a function of pressure for system with FP.D. Operating conditions: O_2/CH_4=0.48, H_2O/CH_4=1.15, SG/CH_4=0

It should be noted that, in FP.D, the maximum value of η (37.2%) is even lower than what is obtained with the conventional ATR reactor (η = 38.5%). This should be attributed to the fact that, notwithstanding the absence of the AOG stream, the dilution of the reacting mixture with nitrogen reduces HR (affecting, in turn, also x_{CH4}) leading to a retentate with relatively high amount of methane and hydrogen, whose heating value cannot be totally recovered.
It should be pointed out that, due to the exothermic nature of the reactions, no additional methane to the burner is required, i.e. α=0, and the exhaust gas stream leaves the plant at quite high temperatures. Data are reported in Table 6.

P (atm)	x_{CH4}	α	T_{EX} (°C)	HR	fa	f_R	η
3.0	85.2	0.0	1369.1	5.8	1.6	60.1	0.5
5.0	88.4	0.0	1248.1	60.2	2.6	67.6	21.8
7.0	90.0	0.0	1178.1	75.6	3.3	71.4	29.1
9.0	90.9	0.0	1132.7	82.7	3.9	73.8	32.7
12.0	91.8	0.0	1089.8	88.0	4.6	76.2	35.6
15.0	92.4	0.0	1063.6	90.9	5.2	77.8	37.2

Table 6. System with FP.C. Operating conditions: O_2/CH_4=0.48, H_2O/CH_4=1.15, SG/CH_4=0

Fig. 12 reports the energy efficiency of system with FP.D as a function of SG/CH_4 parametric in pressure. Simulation details for P = 10 atm are reported in Table 7.

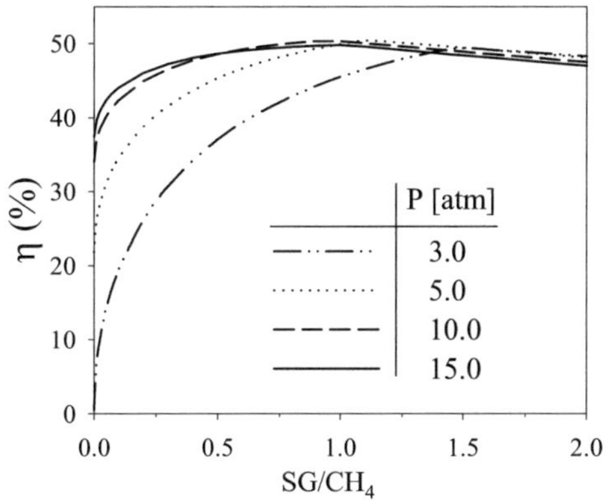

Fig. 12. η as a function SG/CH_4 parametric in pressure for system with FP.D. Operating conditions: O_2/CH_4=0.48, H_2O/CH_4=1.15

SG/CH_4	x_{CH4}	α	T_{EX} (°C)	HR	fa	f_R	η
0.0	91.3	0.0	1115.9	84.9	4.1	74.8	34.0
0.1	95.1	0.0	894.7	95	4.1	81.5	42.4
0.5	99.1	0.0	463.8	99.1	4.1	88.6	48.6
1.0	100	0.4	100.0	99.9	4.1	91.2	50.3
1.5	100	4.0	100.0	100.0	4.1	91.8	48.9
2.0	100	7.0	100.0	100.0	4.1	91.9	47.5

Table 7. System with FP.D. Operating conditions: O_2/CH_4=0.48, H_2O/CH_4=1.15, P=10 atm

The trend of η with SG/CH_4 and pressure is similar to the one observed for the system based on SR. However, it is important to note that for each pressure value investigated, the SG/CH_4 value that maximizes energy efficiency is higher than the corresponding one in the SR-based fuel processor.

This is due to the fact that in an ATR-based system, there is an excess energy due to the autothermic nature of the process, that allows a consistent sweep gas production without methane addition to the burner, i.e. α=0.

Moreover, it is worth noting that energy efficiency of FP.D is highly improved by adding sweep gas, increasing from 34.0% (SG/CH_4=0) to 50.3% (SG/CH_4=1.0).

Table 8 report the detail of the simulation results and value of the operating parameters given as simulation input that maximize the energy efficiency η, for FP.D.

Simulation results						
	f_R	α	HR	f_a	η	T_{EX} (°C)
FP.D	90.2	0.0	99.6	3.3	50.6	100.0
Simulation Input						
		P (atm)	H_2O/CH_4	O_2/CH_4	SG/CH_4	
FP.D (ATR)		7	1.2	0.5	1.0	

Table 8. Membrane ATR - PEMFC system based on membrane reforming reactor.

The best way to operate an autothermal reforming membrane system is to moderately increase pressure and to employ some sweep gas to improve HR (the maximum η is reached for P=7 atm and SG/CH_4=1.0, as reported in Table 7).

The lower value of pressure that maximize η with respect to SR system is due to the higher power required by the auxiliary units, needed essentially to compress the air in the feed.

Finally, it should be noted that the addition of sweep gas in system with FP.D allows reaching energy efficiency values significantly higher than the optimum value of the conventional system (38.5%) and similar to the energy efficiency of SR based systems.

It should be kept in mind that, due to limited thermal stability of the highly selective membranes, membrane units should not be exposed to temperatures higher than 600°C. While FP.C always meets this constraint (since reactor temperature is fixed at 600°C), FP. D does not. Indeed, in the optimal conditions, the first reactors reach temperatures as high as 720°C. Therefore, the actual realization of an integrated membrane reactor would require significant improvements of membrane compatibility with high temperatures. A more realistic configuration of an ATR based membrane reactor should consider a first ATR reactor, where most of the methane oxidation takes place, followed by a membrane reactor, interposing between the two units a heat exchanger to cool down the temperature before entrance into the membrane reactor, so that the membranes are never exposed to temperatures higher than 600°C. With this configuration, energy efficiency becomes 48.5% and the best operating conditions are P=7 atm; O_2/CH_4=0.5; H_2O/CH_4=1.7; SG/CH_4=1.0.

Fuel Processors based on membrane WGS reactor

Optimization performed for systems based on membrane WGS reactors (FP.E for SR and FP.F for ATR) followed the same criteria of what reported for systems based on membrane reforming reactors. Although quantitatively different, the trend of performance with operating parameters were similar to what reported for the systems with membrane reforming reactors, therefore data are not reported for the sake of brevity.

Table 9 reports the simulation results and the value of the operating parameters given as simulation input that maximize the energy efficiency η, for FP.E and for FP.F, respectively.
It is possible to observe that the introduction of the membrane in the WGS reactor allows obtaining higher energy efficiencies than what achieved in the conventional systems.

Simulation results						
	f_R	A	HR	fa	η	T_{EX} (°C)
FP.E (SR)	110.9	18.4	96.8	0.5	52.2	141.5
FP.F (ATR)	83.0	0.0	99.4	1.9	47.6	100.0
Simulation Input						
	P (atm)	H_2O/CH_4	O_2/CH_4	SG/CH_4	T_{SR} (°C)	T_{WGS} (°C)
FP.E (SR)	3	2.0	-	0.2	800	300
FP.F (ATR)	3	1.2	0.6	1.9	-	300

Table 9. Innovative Fuel Processor - PEMFC systems based on membrane WGS reactor

As far as system with FP.E is concerned, the temperature value required for system optimization corresponds to the highest value investigated; this is due to the positive effect of temperature on the SR reactor, and thus on the membrane WGS reactor, that overcomes the negative effect of temperature increase on α.
The maximum efficiency value is limited by the problem of a not complete heat recovery of the exhaust gases (T_{EX}>100°C); this is due to the problem of temperature cross-over that can arise in the heat exchangers when the system works at high SR temperatures.
Since the endothermic nature of the process imposes the necessity of operating with additional methane to the burner, the amount of sweep gas required to optimize the system is small (SG/CH_4=0.2).
It is also possible to observe that the pressure value required for system optimization corresponds to the lowest value investigated; this is due to the negative effect of pressure on the SR reactor, which overcomes the positive effect of pressure increase on the membrane WGS reactor. This one, indeed, allows reaching a high HR, notwithstanding the low pressure value, thanks to the high hydrogen concentration achieved at the outlet of the SR reactor, which positively acts on the driving force.
As far as system with FP.F is concerned, it is possible to observe that the value of H_2O/CH_4 that maximizes the energy efficiency is by far lower than what required for the conventional case. For the ATR systems, indeed, the autothermal nature of the process allows to have an excess heat in the system that can be used to produce steam. In the conventional system, the steam can be used only as reactant, with only moderate improvement of energy efficiency for H_2O/CH_4>3, thus making further steam production useless. In the innovative system, the steam can be used as reactant as well as sweep gas and the energy efficiency resulted to be favored more by an increase of SG/CH_4 than by an increase of H_2O/CH_4.
The autothermal nature of the process allows operating with no additional methane to the burner and the high amount of sweep gas allows the system to operate at low pressure values, favoring the conditions in the ATR reactor.
Although working at the same pressure, the fraction of inlet methane required to run the auxiliary unit is higher in the ATR case than in the SR case, for the presence of air in the feed (f_a=0.5 for FP.E and 1.9 for FP.F).
It is also possible to note that the introduction of the membrane in the WGS reactor not only allows to reach efficiency values higher than what achieved in the conventional systems, but

also makes the SR and ATR based systems similar in terms of energy efficiency (the difference between SR and ATR in the conventional case is ca. 20%, whereas in this case it is only ca. 8%).

5. Conclusions

As a general conclusion on system analysis, the optimum of each fuel processor - PEMFC system and the corresponding operating parameters are reported in Table 9.
It is possible to observe that the SR-based processes always show higher energy efficiency than the corresponding ATR-based processes, with a marked difference in the case of conventional systems (FP.A and FP.B have a difference of about 21% in the energy efficiency value). However, the introduction of the membrane allows to obtain energy efficiency values of the ATR system closer to the efficiency levels reached in the SR ones (differences between SR and ATR based systems of ca. 7% when the membrane is introduced in the reforming reactor and of ca. 9% when the membrane is introduced in the WGS reactor).

	Case	H_2O/CH_4	O_2/CH_4	T_{SR} [°C]	SG/CH_4	P [atm]	η %
SR	FP.A	2.5	-	670	-	1	48.0
	FP.C	2.5	-	600	0.1	15	52.1
	FP.E	2.0	-	800	0.2	3	52.2
ATR	FP.B	4.0	0.56	-	-	1	38.5
	FP.D	1.7	0.5	-	1.0	7	48.5
	FP.F	1.2	0.6	-	1.9	3	47.6

Table 10. Comparison of various FP - PEMFC systems in correspondence of operating conditions that maximize system performance

The comparison between the steam reforming based systems (innovative systems with FP.C and FP.E vs conventional system with FP.A) showed that the employment of a membrane reactor can increase system efficiency from 48.0% to values above 52.0%. Such an efficiency increase requires almost no addition of sweep gas due to the endothermic nature of the process.
The pressure that optimizes the energy efficiency of the two membrane-based system is different; the system with integrated reforming reactor (FP.C) requires to operate at high pressure value (15 atm), whereas the system with membrane WGS reactor (FP.E) at low pressure value (3 atm). This is due to the fact that the SR reactor is negatively influenced by the pressure increase; therefore the system is optimized by increasing the hydrogen recovery in the membrane WGS reactor by increasing hydrogen concentration at the inlet of the WGS reactor more than by increasing pressure.
As regards temperature, all systems require to operate at the highest possible temperature compatible with material stability.
However, although the limit on temperature imposed to the system with membrane reforming reactor is more tighten, energy efficiency results to be as high as the value reached in the system with membrane WGS reactor, that operates at high SR temperature. This is due to the fact than the hydrogen removal from the reaction environment allows to achieve higher performance at lower temperature.
The comparison between the autothermal reforming systems (innovative systems with FP.D and FP.F vs conventional system FP.B) shows that energy efficiency can be improved from

38.5% to values around 48%, if a membrane reactor is employed. To obtain such an energy efficiency improvement, sweep gas addition is required.
The considerations on pressure are the same of what reported for the SR case, although the system with membrane reforming reactor is optimized at pressure values lower that the SR case (7 atm instead of 15 atm) due to the higher value of power required to run the auxiliary units.
It is possible to observe that the value of H_2O/CH_4 that maximizes the energy efficiency of the innovative ATR systems is far lower than what required for the conventional case.
Indeed, in the innovative systems, the steam can be used as reactant and as sweep gas and the energy efficiency resulted to be favored more by an increase of SG/CH_4 than by an increase of H_2O/CH_4.

6. References

Ahmed, S. Krumpelt, M. (2001). Hydrogen from hydrocarbon fuels for fuel cells, *International Journal of Hydrogen Energy,* 26, 291–230, ISSN 0360-3199

Aspen Technology Inc. http://www.aspentech.com

Basile, A. Chiappetta, G. Tosti, S. Violante, V. (2001). Experimental and simulation of both Pd and Pd/Ag for a water gas shift membrane reactor, *Separation and purification technology,* 25, 549-571, ISSN 1383-5866

Campanari, S. Macchi, E. Manzolini, G. (2008). Innovative membrane reformer for hydrogen production applied to PEM micro-cogeneration: Simulation model and thermodynamic analysis, *International Journal of Hydrogen Energy,* 33, 1361-1373, ISSN 0360-3199

Ersoz, A. Olgun, H. Ozdogan, S. (2006). Reforming options for hydrogen production from fossil fuels for PEM fuel cells, *Journal of Power Sources,* 154, 67-73, ISSN 0378-7753

Francesconi, JA. Mussati, MC. Mato, RO. Aguirre, PA. (2007). Analysis of the Energy efficiency of a integrated ethanol processor for PEM fuel cell systems, *Journal of Power Sources,* 167, 151–161, ISSN 0378-7753

Hou, J. Zhuang, M. Wag, G. (2007). The analysis for the efficiency properties of the fuel cell engine, *Renewable Energy,* 32, 1175-1186, ISSN 0960-1481

Larminie, J. & Dicks, A. (2004). *Fuel Cell Systems Explained, Second Edition,* John Wiley & Sons Ltd, 0-470-84857-X, The Atrium, Southern Gate, Chichester, West Sussex PO19 8SQ, England

Lampert, J. (2004). Selective catalytic oxidation: a new catalytic approach to the desulfurization of natural gas and liquid petroleum gas for fuel cell reformer applications, Journal of Power Sorces, 131, 27-34, ISSN 0378-7753

Lattner, R. Harold, MP. (2004). Comparison of methanol-based fuel processors for PEM fuel cell systems. *International Journal of Hydrogen Energy,* 29, 393-417, ISSN 0360-3199

Lattner, R. Harold, MP. (2005). Comparison of conventional and membrane reactor fuel processors for hydrocarbon-based PEM fuel cell systems, *Applied Catalysis B: Environmental,* 56, 149-169, ISSN 0926-3373

Lyubovsky, M. Walsh, D. (2006). Reforming system for co-generation of hydrogen and mechanical work, *Journal of Power Sources,* 157, 430–437, ISSN 0378-7753

Manzolini, G. Tosti, S. (2008). Hydrogen production from ethanol steam reforming: energy efficiency analysis of traditional and membrane processes, *International Journal of Hydrogen Energy,* 33, 5571-5582, ISSN 0360-3199

Pearlman, B. Bhargav, A. Shields, EB. Jackson, GS. Hearn, PL. (2008). Modeling efficiency and water balance in PEM fuel cell systems with liquid fuel processing and hydrogen membranes, *Journal of Power Sources,* 185, 1056–1065, ISSN 0378-7753

Ratnamala, GM. Shah, N. Mehta, V. Rao, PV. (2005). Integrated Fuel Cell Processor for a 5-kW Proton-Exchange Membrane Fuel Cell, *Industrial & Engineering Chemistry Research,* 44, 1535-1541, ISSN 0888-5885

Semelsberger, TA. Brown, LF. Borup, RL. Inbody, MA. (2004). Equilibrium products from autothermal processes for generating hydrogen-rich fuel-cell feeds, *International Journal of Hydrogen Energy,* 29, 1047-1064, ISSN 0360-3199

Seo, YS. Shirley, A. Kolaczkowski, ST. (2002). Evaluation of thermodynamically favourable operating conditions for production of hydrogen in three different reforming technologies, *Journal of Power Sources,* 108, 213–225, ISSN 0378-7753

Shu, J. Grandjean, BPA. van Neste, A. Kaliaguine, S. (1991). Catalytic palladium-based membrane reactors: a review, *Canadian Journal of Chemical Engineering,* 69, 1036-1060, ISSN 0008-4034

Song, C. (2002). Fuel processing for low-temperature and high-temperature fuel cells: Challenges, and opportunities for sustainable development in the 21st century, *Catalysis Today,* 77, 17–49, ISSN 0920-5861

Xu, J. Froment, GF. (1989). Methane Steam Reforming, Methanation and Water-Gas Shift. Intrinsic Kinetics, *AIChE Journal,* 35, 88-96, ISSN 0001-1541